MECHANICAL ENGINEERING THEORY AND APPLICATIONS

ROBUST CONTROL

SYSTEMS, THEORY AND ANALYSIS

MECHANICAL ENGINEERING THEORY AND APPLICATIONS

Additional books in this series can be found on Nova's website under the Series tab.

Additional e-books in this series can be found on Nova's website under the eBooks tab.

Mechanical Engineering Theory and Applications

Robust Control

Systems, Theory and Analysis

Sharon Bennett

Editor

nova publishers

New York

Library of Congress Cataloging-in-Publication Data

Names: Bennett, Sharon (Engineer), editor.
Title: Robust control : systems, theory and analysis / editor, Sharon Bennett.
Other titles: Robust control (Bennett)
Description: Hauppauge, New York : Nova Science Publishers, Inc., [2017] |
Series: Mechanical engineering theory and applications | Includes
bibliographical references and index.
Identifiers: LCCN 2016059870 (print) | LCCN 2017000232 (ebook) | ISBN
9781536108262 (hardcover) | ISBN 9781536108477 (Ebook) | ISBN
9781536108477
Subjects: LCSH: Electric controllers--Reliability. | Control theory.
Classification: LCC TK2851 .R585 2017 (print) | LCC TK2851 (ebook) | DDC
629.8--dc23
LC record available at https://lccn.loc.gov/2016059870

Published by Nova Science Publishers, Inc. † New York

CONTENTS

PREFACE

The Robust Control Theory involves powerful methods for analysis and design of control systems in presence of signal and parameter uncertainties. The most frequently used techniques for robust control design are the H_∞ design and the μ–synthesis. In this book, Chapter One reviews issues related to the design and practical implementation of high order robust controllers. Chapter Two deals with multi-objective disturbance attenuation control and filtering problems for disturbances from different classes. Chapter Three discusses a robust control design for general switched affine control systems. Chapter Four presents a PID control scheme for a synchronous motor with permanent magnets.

Chapter 1 is devoted to various issues related to the design and practical implementation of high order robust controllers. The authors consider derivation of plant uncertainty models using analytical or identification procedures, implementation of different schemes for μ-synthesis, choice of weighting filters and controller order reduction. Additional important problems arising in the framework of embedded control systems, like removing the sensor drifts, generation of control code from Simulink® and effect of single precision arithmetic on the controller stability, are discussed in some details. As a case study the authors present the robust control of two-wheeled robot using μ-controllers of order up to 30.

A multi-objective disturbance attenuation problem is considered in Chapter 2 as a framework for control and filtering problems under multiple exogenous disturbances and unknown initial conditions. Necessary conditions of the Pareto optimality in this problem are derived. Pareto suboptimal solutions are shown to be optimal solutions with respect to a multi-objective cost which is a parameterized disturbance attenuation level of the system under disturbances acting on all channels simultaneously. It is established that the relative losses of these Pareto suboptimal solutions compared to the Pareto optimal ones do not exceed $1-\sqrt{N}/N$, where N is the number of criteria. This approach is applied to control and filtering problems in continuous-time and discrete-time systems with multiple classes of L_2-bounded and impulsive disturbances for which the disturbance attenuation levels are H_∞ and γ_0 norms. The H_∞/γ_0 optimal controllers and filters as the Pareto suboptimal solutions are synthesized in terms of linear matrix inequalities (LMIs). All these results are extended to the case when the disturbances acting on different inputs are combined into coalitions.

Chapter 3 deals with a robust control design for general switched affine control systems. Dynamical models under consideration are described by ordinary differential equations involving a switching mechanism and in the presence of bounded uncertainties. The design procedure the authors analyse is based on the newly elaborated attractive ellipsoids method. The stability and robustness of the resulting closed-loop system involves an abstract Clarke stability theorem and a theoretic extension of the celebrated Lyapunov-type methodology. A short discussion on the obtained analytic results and possible applications and extensions is also included.

Chapter 4 presented a PID control scheme for a synchronous motor with permanent magnets. PID controllers are the most used controllers in industrial applications and it always presents difficulty to realise the differential part. Moreover, in the last decades a visible effort has been made in the field of sensorless control in which observer and approximation of state variables are used to avoid sensors in the control loop. A polynomial Kalman filter is used to approximate its derivative actions to avoid disturbances and spikes due to the differentiation. The Kalman filter estimates two coefficients at each sampling time, which are necessary to calculate the polynomial of the first order with which the tracking error is intended to be modelled. Simulation results are shown to demonstrate the effectiveness of the approximation.

In: Robust Control: Systems, Theory and Analysis ISBN: 978-1-53610-826-2
Editor: Sharon Bennett

Chapter 1

DESIGN AND IMPLEMENTATION OF ROBUST CONTROLLERS: THEORETICAL AND NUMERICAL ISSUES

Ts. Slavov, J. Kralev and P. Petkov*
Department of Systems and Control,
Technical University of Sofia, Sofia, Bulgaria

Abstract

This Chapter is devoted to various issues related to the design and practical implementation of high order robust controllers. We consider derivation of plant uncertainty models using analytical or identification procedures, implementation of different schemes for μ-synthesis, choice of weighting filters and controller order reduction. Additional important problems arising in the framework of embedded control systems, like removing the sensor drifts, generation of control code from Simulink® and effect of single precision arithmetic on the controller stability, are discussed in some details. As a case study we present the robust control of two-wheeled robot using μ-controllers of order up to 30.

Keywords: robust control, real time control, embedded systems, Digital Signal Processors

AMS Subject Classification: 93B35, 93C83, 93D09, 93E12

*E-mail address: php@tu-sofia.bg

1. Introduction

The Robust Control Theory involves powerful methods for analysis and design of control systems in presence of signal and parameter uncertainties (Sánchez-Peña and Sznaier 1998; Skogestad and Postlethwaite 2005; Zhou et al. 1996). The most frequently used techniques for robust control design are the $\mathcal{H}_\infty$ design and the μ–synthesis (Skogestad and Postlethwaite 2005). The $\mathcal{H}_\infty$ optimization is usually preferred in robust design because it produces controllers of smaller order which facilitates their implementation. The common disadvantage of all $\mathcal{H}_\infty$ design methods is that they are suitable for plants with unstructured uncertainties but can not ensure robust stability and robust performance in the general case of unstructured and structured (parametric) uncertainties. In contrast, the μ–synthesis which aims at minimization of the structured singular value (Zhou et al. 1996) may ensure robust stability and robust performance in the presence of exogenous disturbances, noises and different type of uncertainties. The high order of the controller obtained is usually pointed out as a disadvantage of μ–synthesis. However, with the appearing of powerful and cheap processors in the recent years this peculiarity of the μ–synthesis does not pose a significant difficulty.

In contrast with the theoretical achievements, the practical implementation of robust control laws is still in its beginning. There is a few real life applications of high order robust control laws reported in the literature (see for instance (Bautista-Quintero and Pont 2008; Howlader et al. 2013; Raafat et al. 2012). The main obstacle of robust control laws implementation is the difficulties related to the development, testing and verification of the necessary real-time software which is highly dependent on the type of digital controller platform used. These difficulties are reduced significantly using the recent technologies for automatic code generation and embedding implemented in MATLAB®/Simulink®program environment (Simulink Coder 2013).

This Chapter is devoted to various issues related to the design and practical implementation of high order robust control laws. We consider derivation of plant uncertainty models using analytical or identification procedures, implementation of different schemes for μ-synthesis, choice of weighting filters and controller order reduction. Additional important problems arising in the framework of embedded control systems, like removing the sensor drifts, generation of control code from Simulink®and effect of single precision arithmetic on the controller stability, are discussed in some details. As a case study we present the robust control of two-wheeled robot using μ-controllers of order up to 30.

The Chapter is organized as follows. First, we consider the derivation of uncertain plant models using analytical models or identification procedures. We illustrate the difficulties related to the identification of uncertain models by an example. In the next Section we give a brief overview of Robust Control Theory based on the usage of structured singular value μ. Several implementation aspects of μ-synthesis are discussed including weighting filters selection and D-K-iterations convergence. The third Section is devoted to the implementation of robust controllers in embedded control systems. The issues discussed involve removing the sensor drifts and effect of single precision arithmetic on the controller stability. In the final Section we present as a case study the design and implementation of robust controller intended for stabilization of two-wheeled self-balancing robot. The experimental results confirm that the closed-loop system achieves both robust stability and robust performance in respect to the uncertainties related to the identification of robot model.

2. Derivation of Uncertain Plant Models

As noted in (Bittanti and Garatti 2012), control science is basically a model based discipline and the performance of control is determined by the accuracy of the model representing data. Building uncertain system models is an important step in the design of robust control systems. Unfortunately, the derivation of uncertain plant models may be much more difficult in comparison with modeling of plants with negligible uncertainties.

2.1. Implementation of Analytical Models

In practice, it is preferable to have an analytical nonlinear plant dynamics model which is used in the derivation of linearized uncertainty model. The analytical models allow to obtain easily both structured (parametric) and unstructured (complex) uncertainty models. Also, the analytical description may be used to obtain the so-called *parameter-dependent model* (Gahinet et al. 95, Ch. 2) which depends on parameters that may undergo large variations along the time.

Implementing the nonlinear model in Simulink®, it is possible to use the corresponding functions of Robust Control Toolbox®3 to linearize the plant and extract an uncertainty model represented in Figure 1. The linearized uncer-

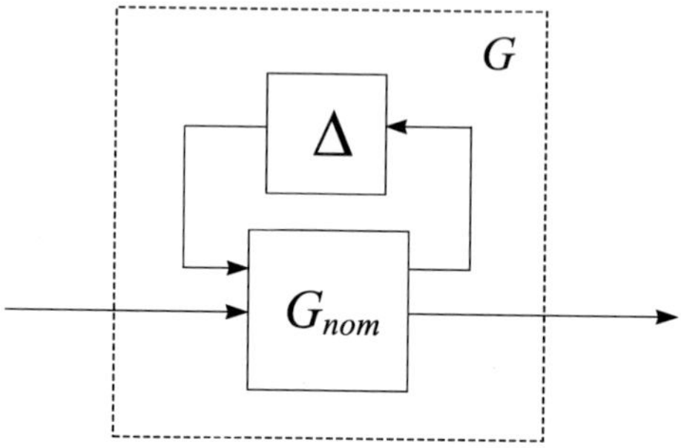

Figure 1. Uncertain plant.

tain plant model G is in the form of Linear Fractional Transformation (LFT), where G_{nom} is the transfer function matrix of the nominal model and Δ is a block-diagonal uncertainty matrix. The Robust Control Toolbox®3 functions also allow to obtain the time and frequency response characteristics of the uncertainty plant models.

2.2. Implementation of Identification Procedures

Frequently, a reliable analytical model of the plant dynamics is not available and the system designer is faced with the necessity to identify the model from experimental data. As it is well known (Isermann and Münchhof 2011; Landau and Zito 2006; Ljung 1987) identification from noisy data may represent a difficult task especially in case of unstable plant when the measurements are obtained in closed-loop.

The following example illustrates the difficulties in deriving uncertainty model of an unstable plant using the functions of System Identification Toolbox®3 [20].

The input-output data obtained by closed-loop experiment for a single-input, single-output system is shown in Figure 2. It consists of 3000 measurements with sampling period of 0.005 s, additional 200 data samples being used for model validation. Based on the three criteria - model loss function, Akaike information index and Rissanen index - a fourth order Box-Jenkins model is chosen for further identification. It is done by the function `bj` using data sets of 1000, 2000 and 3000 data samples. In all three cases the models obtained pass through the test of autocorrelation and crosscorrelation function of model

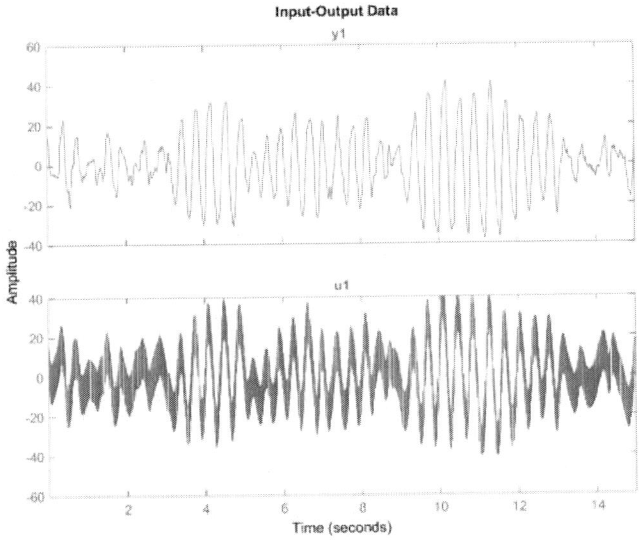

Figure 2. Input-output data used in the identification.

residuals, which indicates unbiased parameter estimates. The latter condition guarantees that the exact parameter values are contained in the confidence intervals of estimates with probability close to 1. Based on confidence intervals, maximum relative deviations from nominal models are obtained, which may be used to derive models with unstructured uncertainty. In Figures 3 - 8 we show bounds on the relative uncertainties derived for the three models along with the Bode plots of the corresponding uncertain plant models.

It is seen from the Figures that the picks of the relative uncertainties derived vary between 1.3 for the case of 2000 samples and 4.3 for the case of 1000 samples. Clearly, the uncertainty model obtained for 1000 data samples is not appropriate for controller design. So different results for the uncertainty bounds show that additional optimization is necessary in order to obtain a model with minimum of uncertainty bound. The alternative is to use special identification methods for uncertainty models, see for instance (Gugercin et al. 2003; Van den Hof 2001; Venkatesh 2003).

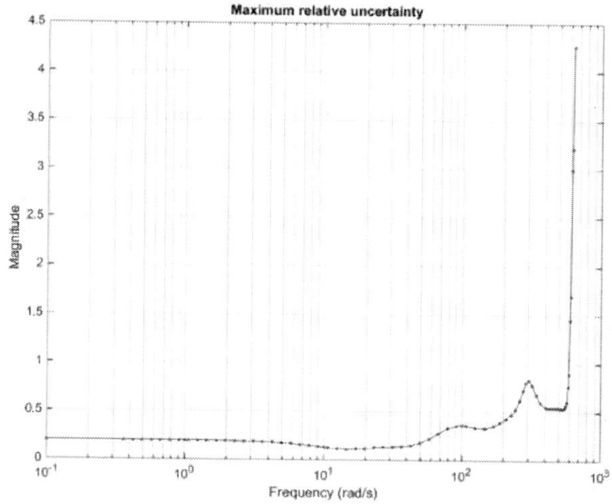

Figure 3. Relative uncertainty for 1000 samples.

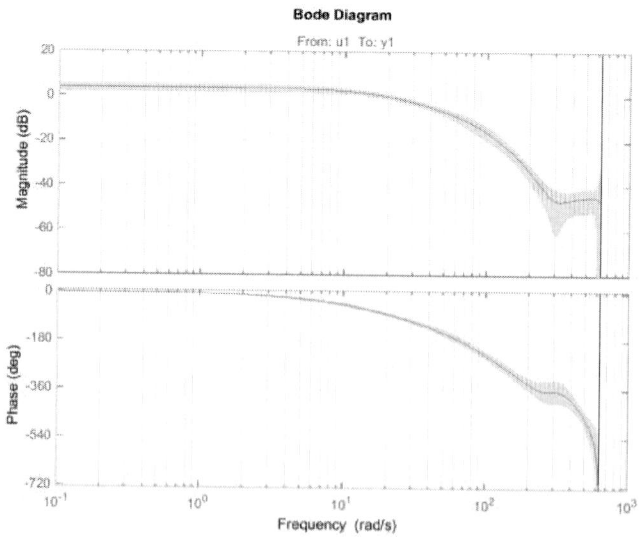

Figure 4. Uncertain model for 1000 samples.

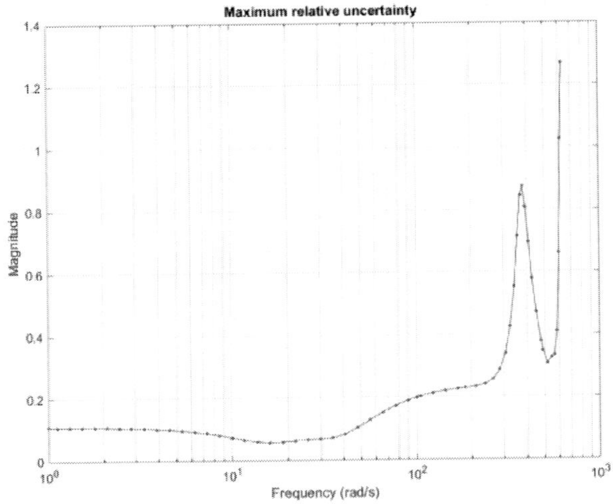

Figure 5. Relative uncertainty for 2000 samples.

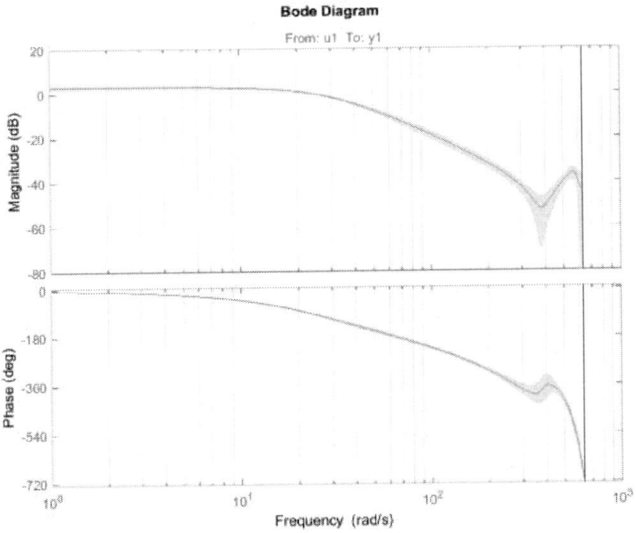

Figure 6. Uncertain model for 2000 samples.

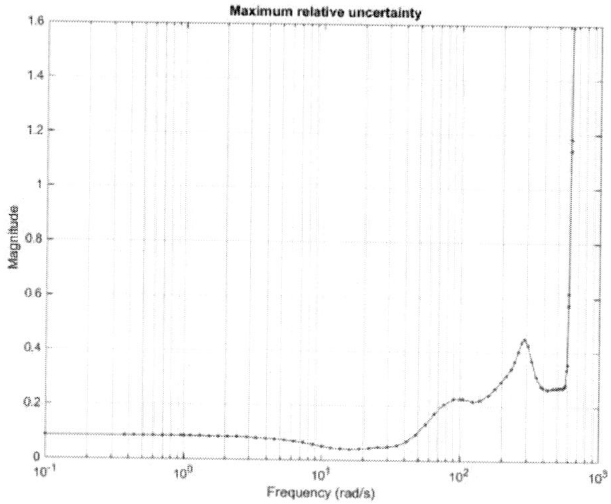

Figure 7. Relative uncertainty for 3000 samples.

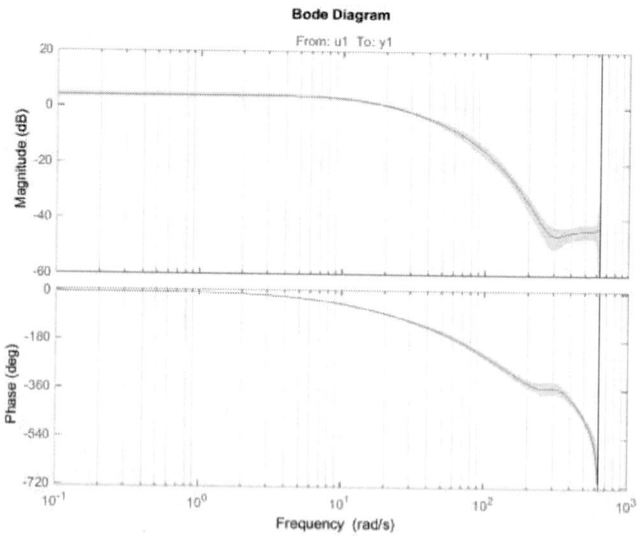

Figure 8. Uncertain model for 3000 samples.

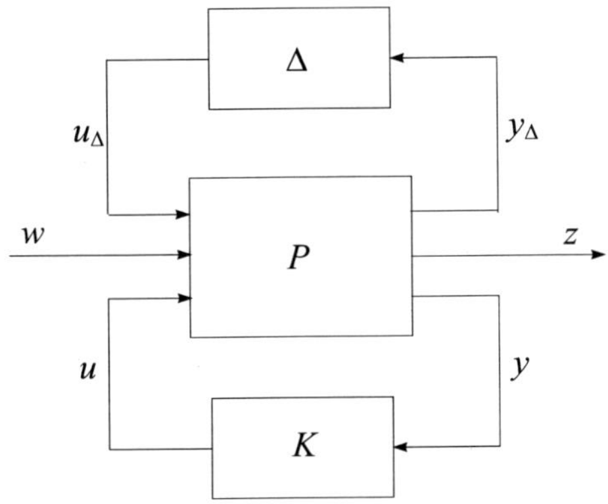

Figure 9. Uncertain closed-loop system.

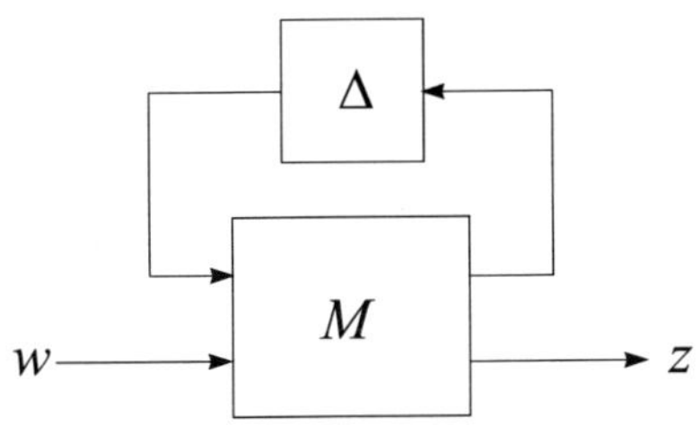

Figure 10. $M - \Delta$ loop for robust stability analysis.

3. Robust Control Design

Consider a control problem in the Linear Fractional Transformation shown in Figure 9.

The system denoted by P is the open-loop connection and represents all known elements including the nominal system model and the performance weighting functions, as well as the uncertainty weighting functions. The block Δ parameterizes all supposed model uncertainty. The controller is denoted by K. Inputs to P are three sets of signals: inputs u_Δ due to the uncertainty, references and disturbances w and controls u. Three sets of outputs are generated: outputs y_Δ due to the uncertainty, errors or controlled outputs z and measurements y.

For the aim of robust stability analysis, it is convenient to represent the uncertain control system by the $M-\Delta$ loop, shown in Figure 10. In this loop the transfer function of the nominal part (denoted by M) is separated from uncertain part (denoted by Δ). We have that $M = F_\ell(P,K)$ where $F_\ell(P,K)$ denotes the transfer function matrix of the lower Linear Fractional Transformation of P and K. It is possible to prove that the system shown in Figure 10 will achieve robust stability for all Δ if and only if

$$\mu_\Delta[M(j\omega)] < 1 \tag{1}$$

where $\mu_\Delta(F_\ell(P,K))$ is the *structured singular value* of the closed-loop system (Zhou et al., 1996). The design goal is to determine a controller K, stabilizing the nominal system as well as for all $\Delta, \max_\omega \bar{\sigma}[\Delta(j\omega)] \leq 1$. The closed-loop system achieves robust performance if the system is stable and satisfies the performance criterion

$$\|F_U[M,\Delta]\|_\infty < 1 \tag{2}$$

where $F_U(M,\Delta)$ is the transfer function matrix of the upper LFT of M and Δ. For given arbitrary K this criterion may be tested by using the robust performance test on the nominal part $M = F_\ell(P,K)$. The robust performance test should be performed in respect to the extended uncertain structure

$$\Delta_P \stackrel{\text{def}}{=} \left\{ \begin{bmatrix} \Delta & 0 \\ 0 & \Delta_F \end{bmatrix} : \Delta_F \in \mathbb{C}^{n_w \times n_z} \right\}, \tag{3}$$

where Δ_F is a fictitious complex (unstructured) uncertainty. The system with controller K achieves robust performance if and only if

$$\mu_{\boldsymbol{\Delta_P}}(F_\ell(P,K)(j\omega)) < 1 \tag{4}$$

where $\mu_{\boldsymbol{\Delta_P}}(F_\ell(P,K)$ is structured singular value determined with respect to the extended uncertainty Δ_P.

In the case of $\mathcal{H}_\infty$-design we look for a controller which ensures

$$\|M(s)\|_\infty < \gamma \tag{5}$$

for a small positive number γ. Note that the system uncertainty is not taken into account in the $\mathcal{H}_\infty$-design.

The main disadvantage of the $\mathcal{H}_\infty$-design is that it does not allways ensure robust stability and robust performance of the closed-loop system in presence of uncertainty in the plant.

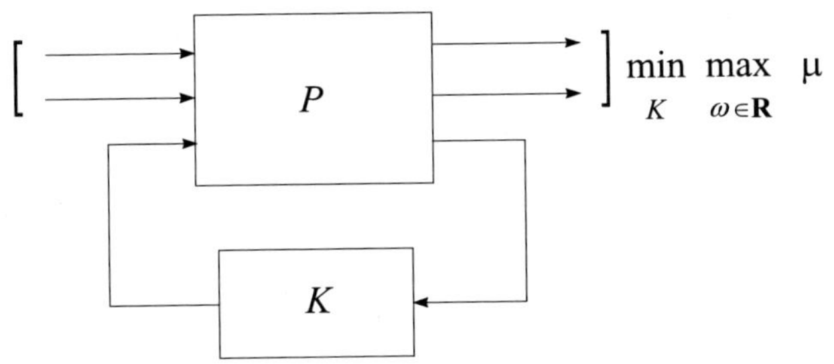

Figure 11. Block-diagram of the closed-loop system.

3.1. μ-Synthesis Problem

The μ–synthesis is one of the most important techniques in Robust Control Design. For properly chosen weighting functions this design method usually produces a controller that ensures both robust stability and robust performance of the closed-loop system.

Consider a control problem shown in Figure 11.

The aim of the *μ–synthesis* is to minimize the peak value of the structured singular value $\mu_{\mathbf{\Delta_P}}(.)$ of the closed-loop transfer function matrix $F_\ell(P, K)$ over the set of all stabilizing controllers K. This is written as

$$\min_{\substack{K \\ \text{stabilizing}}} \max_{\omega} \mu_{\mathbf{\Delta_P}}(F_\ell(P, K)(j\omega)). \tag{6}$$

The μ-synthesis has the following features.

1. Good disturbance attenuation and noise suppression.
2. Achieves stability and performance robustness.
3. Doesn't require accurate plant model.
4. Difficult tuning of the weighting functions.
5. The controller order is higher than the order of the $\mathcal{H}_\infty$ controller.

Consider how to set the performance requirements in the μ-synthesis.

The design goal is to achieve robust stability and robust performance of the closed-loop system shown in Figure 12 in the presence of plant uncertainty and

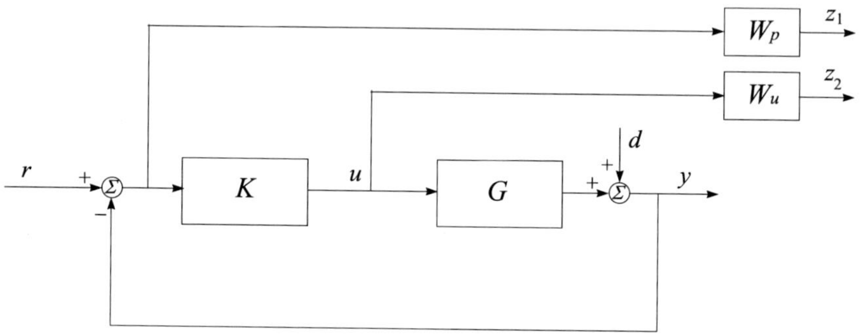

Figure 12. Block-diagram of the closed-loop system.

output disturbances. The matrix transfer functions W_p and W_u are frequency dependent weighting functions (filters) that will be called weighting performance function and weighting control function, respectively. The role of the function W_p is to "penalize" the difference between system output y and reference r in desired low frequency range and the role of the function W_u is to limit the magnitude of control action u especially in the high frequency range.

The closed-loop system is described by

$$z = T_{zw}w, \tag{7}$$

where

$$z = \begin{bmatrix} z_1 \\ z_2 \end{bmatrix}, \quad w = \begin{bmatrix} r \\ d \end{bmatrix},$$

$$T_{zw} = \begin{bmatrix} W_pS_o & -W_pS_o \\ W_uKS_o & -W_uKS_o \end{bmatrix},$$

and

$$S_o = (I - GK)^{-1}$$

is the output sensitivity transfer function matrix.

To achieve closed-loop robust performance, i.e.,

$$\mu_{\boldsymbol{\Delta_P}}(T_{zw}(j\omega)) < 1, \tag{8}$$

means that the condition

$$\|T_{zw}\|_\infty < 1 \tag{9}$$

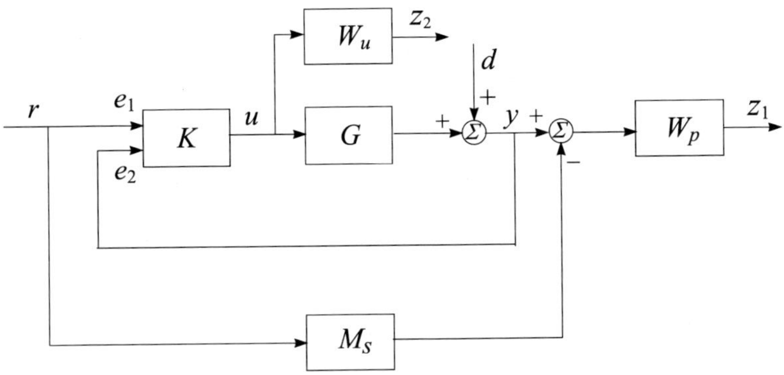

Figure 13. Design of two-degree-of-freedom controller.

will be fulfilled for each possible plant uncertainty. This, in turn, guarantees fulfillment of the conditions

$$\|W_p S_o\|_\infty < 1, \quad \|W_u K S_o\|_\infty < 1. \tag{10}$$

Figure 13 illustrates the implementation of the so called *two-degree-of-freedom controller*. The goal in this case is to achieve closeness of the closed-loop system behavior to the model M_s for all possible plant uncertainties. The usage of two-degree-of-freedom controller usually lead to better design results in comparison with the usage of one-degree-of-freedom controller.

3.2. Choice of Weighting Filters

The performance weighting transfer functions W_p are chosen as low pass filters to suppress the output disturbance d, and the control weighting transfer functions W_u are chosen as high pass filters with appropriate bandwidth in order to impose constraints on the high frequency spectrum of the control actions [11]. This is based on the assumption that the disturbance spectrum is in the low-frequency range and the noise spectrum is in the high frequency range. In some cases it is necessary to modify the weighting functions many (hundred and more) times in order to achieve the desired closed-loop behavior which is one of the few disadvantages of the μ-synthesis. The choice of weighting functions may be done by implementing optimization procedures.

3.3. Implementation of D-K Iterations

The μ synthesis is done via the so called D-K iterations (Zhou et al. 1996) which represent an approximate method for finding the minimum of pick value of μ. These iterations are realized in Robust Control Toolbox®3 by the function `dksyn`(Balas et al. 2013). The convergence of D-K iterations is not guaranteed and in some cases one may observe very poor iteration behavior or even divergence. In such cases it is possible to make slight changes in the performance weighting functions or in the model parameters in order to improve the convergence. Also, the inclusion of small noises in some system inputs may have very good impact on the D-K iterations.

Below we give some recommendations about how to do in practice the μ-synthesis.

1. Due to the efforts necessary to derive the uncertain plant model and the almost unavoidable complication of the controller designed it is appropriate to begin with simplified uncertainty description in order to see wether the performance requirements can be met. Only in the case when these requirements are satisfied, it is appropriate to consider more complicated uncertainty descriptions including, for instance, parametric uncertainties to "sharpen" the design with more accurate uncertainty model.

2. The usage of μ means worst-case analysis so one should be cautious when introducing many sources of uncertainties, disturbances and noises. In such a case it becomes less and less possible for the worst case to appear and the analysis and design performed may become unnecessarily conservative.

3. There is always uncertainty in respect to inputs and outputs so that it is reasonable in the general case to include diagonal input and output uncertainties. The relative (multiplicative) uncertainty is very appropriate for this aim.

4. In the practical design it is customary to obtain values of μ that exceed one. This may result from very high requirements to the closed-loop performance which are impossible to satisfy for the given plant. In such a case it is necessary to loosen the requirements setting other performance and/or control action weighting functions. Finding a controller that ensures value of μ less than one means that the requirements set are pos-

sible to achieve. In other cases it may be necessary to design controller with different structure, for instance two-degree-of-freedom controller, in order to fulfil the requirements posed.

5. In case of a discrete-time system it is appropriate to perform the controller design initially in the continuous-time case. This is due to the fact that the best possible performance may be obtained in the continuous-time case which can then be considered as a limit for discrete-time designs. Also, in the continuous-time case it is easier to find appropriate performance weighting functions that again may be implemented in the discrete-time design.

4. Controller Implementation

4.1. Embedded Control Systems

Implementation of robust controllers in real-time is done by *embedded control systems* which are characterized by the fact that the control software is stored in read-only memory (Hristu-Varsakelis and Levine 2005, Part III). The process of embedding the control software by using MATLAB®and Simulink®is illustrated in Figure 14. The robust control law is designed in MATLAB®and its Simulink®model is translated to the embedded processor using the automatic generation code tools Simulink Coder®(Simulink Coder 2013) and Embedded Coder®(Embedded Coder 2013). The embedded systems used for motion control usually implement Microelectromechanical systems (MEMS) sensors.

Integration of hardware and software in embedded control systems proceeds in three phases: software-in-the-loop, processor-in-the-loop and hardware-in-the-loop. In the first phase the hardware is represented entirely by Simulink®models. The control code is generated automatically from Simulink®model and is validated off-line by simulation of the closed-loop system. At this stage it is possible to use Monte Carlo simulation which can take into account parametric variations and nonlinear effects. In the second phase the control code is tested in the real-time embedded processor using hardware simulated by Simulink®models. Finally, in the third phase the control code is tested with the prototyped hardware in order to verify the integrated functional and operational performance in real environment (strong disturbances, noises and parametric variations). This technique makes possible to examine rapidly

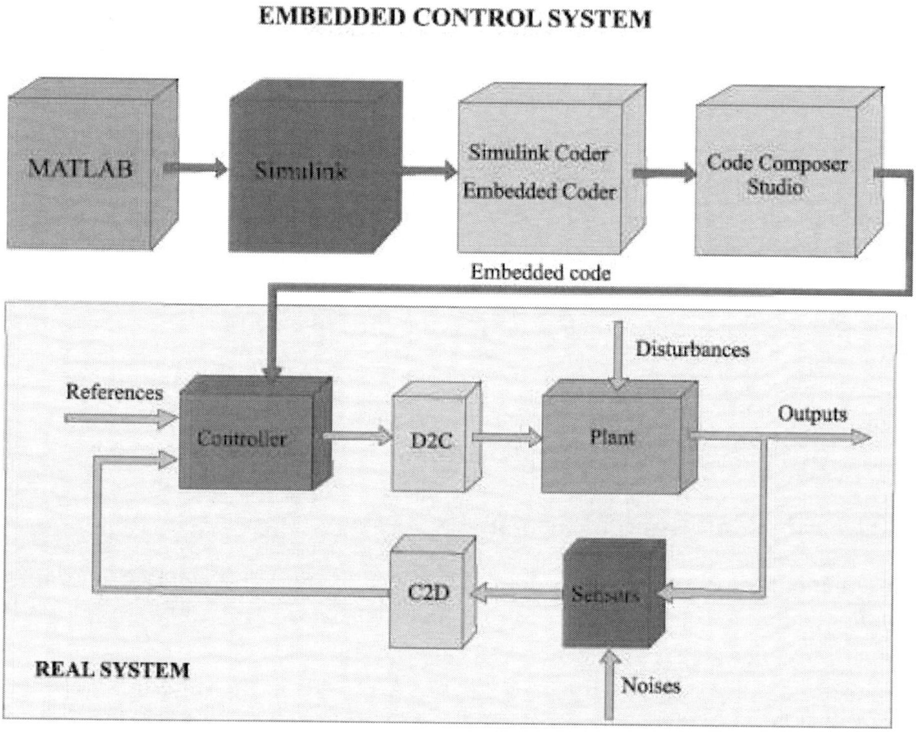

Figure 14. Block-diagram of an embedded control system.

different control laws reducing in the same time the danger of accidents during real experiments.

The embedded control systems have the following distinguishing features.

1. Signals are presented as fixed-point integers or single precision numbers. Floating-point arithmetic may be emulated or implemented by integrated hardware floating-point unit.

2. The controller model should be modified to include I/O interfaces with the external devices.

3. The MEMS sensor noises may be intensive in some frequency ranges.

Analysis of embedded control systems requires the usage of methods pertaining to the theory of hybrid control systems (Lunze and Lamnabhi-Lagarrigue 2012).

4.2. Sensor Errors

The sensor errors may have a strong impact on the closed-loop system performance since these errors are amplified by the controller at the plant inputs.

The MEMS sensor errors consist of deterministic and stochastic parts. The deterministic part includes constant biases, scale factors, axis misalignment and so on, which are removed from row measurements by the corresponding calibration techniques. The stochastic part contains random errors (noises) which cannot be removed from the measurements and should be modeled as stochastic processes.

As an example, consider the MEMS gyroscope used in motion control. Its noise typically consists of the following terms:

- *Bias instability.* This is a stationary stochastic process which may be considered as a low-order zero-mean Gauss-Markov process.
- *Angular random walk.* This is an angular error process which is due to white noise in angular rate.
- *Rate random walk.* This is a rate error due to white noise in angular acceleration.
- *Discretization error.* This is an error representing the quantization noise.

Several other noise terms are described in detail in (IEEEStd 952-1997).

Different techniques for building models of MEMS sensor noises are presented in (El-Sheimy et al. 2008; Quinchia et al. 2012; Petkov and Slavov 2010), to name a few. Usually, they exploit the autocorrelation function of the noise in order to obtain 1st order Gauss-Markov or higher order Auto-Regressive models. Note that it is desirable to keep the model order as low as possible since the model is frequently used in the design of Kalman filter to determine optimal estimates based on the sensor measurements.

Stochastic discrete-time models of MEMS gyro are obtained on the basis of frequency-domain and time-domain characteristics of the sensors noises. For this aim it is possible to use the power spectral density of gyro noise (see Figure 15) and the so called *Alan variance* (Allan 1966) shown in Figure 16. By using

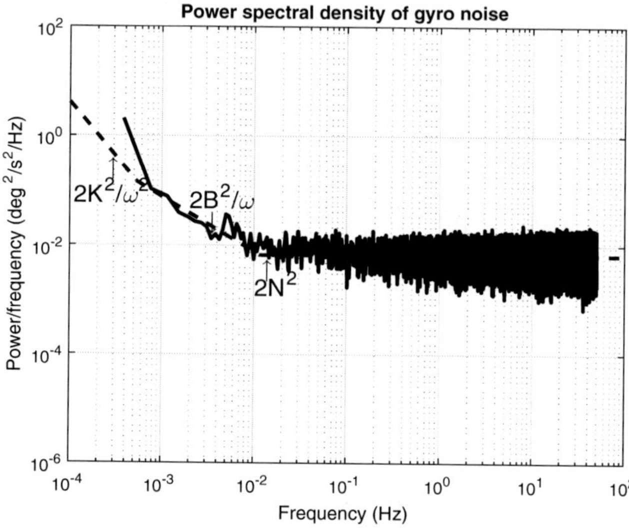

Figure 15. Power spectral density of gyro noise.

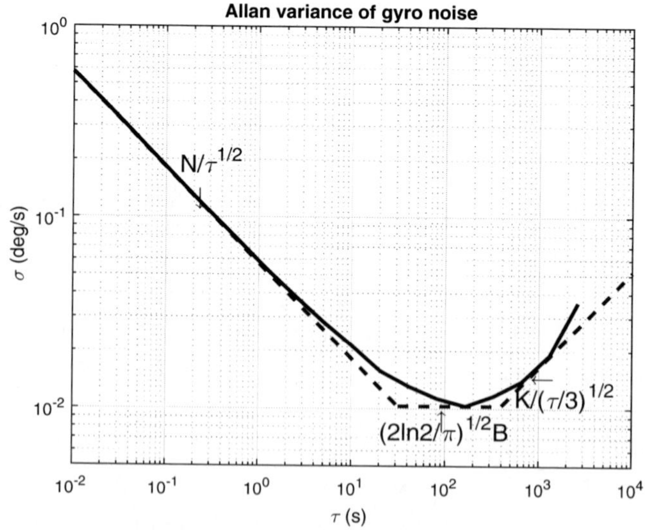

Figure 16. Allan variance of gyro noise.

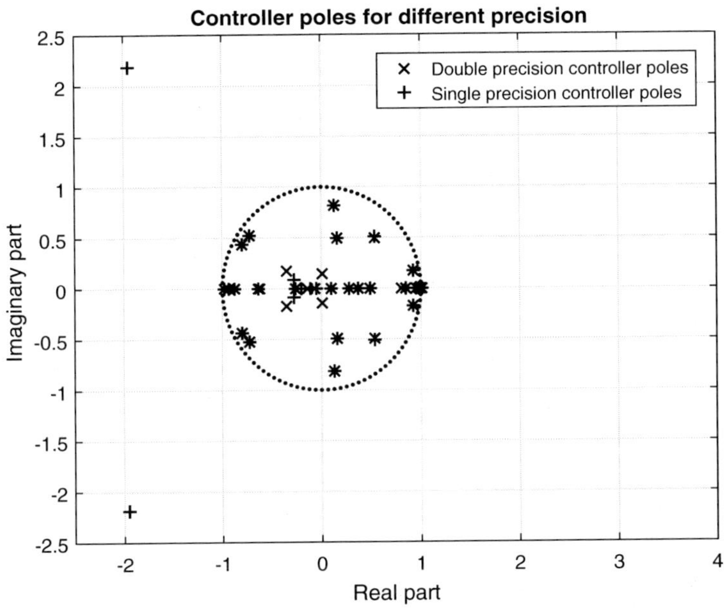

Figure 17. Influence of the precision used on controller poles.

nonlinear least-squares minimization it is possible to determine the numbers B, N and K which characterize bias instability, angular random walk and rate random walk, respectively.

Finally, the discretization error $e(k)$ in sensor measurements and analog-to-digital converters at each sampling instant k satisfies the inequality $-q/2 \leq e(k) \leq q/2$ where the quantum size q is determined by $q = D/2^t$, D is the range of the variable measured and t is the number of bits in floating point representation of the corresponding device output. This error may be described by a white random process with intensity $q^2/12$ and for values of t which are usually less than or equal to 12 may lead to significant noise at the plant input.

4.3. Influence of the Working Precision

The robust control laws are designed in MATLAB®using double precision (64 bits) floating point arithmetic. In the best case, these control laws are implemented in the embedded processor by using single precision (32 bits) arithmetic. This circumstance may affect significantly the behavior of the discrete-time closed-loop system. As an example we show in Figure 17 the poles of a discrete-time robust controller of order 40 implemented in double and single precision. As it is seen from the Figure, the controller implemented in single precision becomes unstable which is undesirable in practice. This shows that controller stability in single precision should be checked after the design in MATLAB®.

5. A Case Study: Robust Control of Two-Wheeled Robot

Two-wheeled robots have several applications which make them interesting from theoretical an practical point of view. The most popular commercial product, built on the idea of self balancing two-wheeled robot is the Segway®Personal Transporter (PT), produced by Segway Inc., USA (Segway 2012). Some of the Segway®PTs have maximum speed of 20 km/h and can travel as far as 38 km on a single battery charge. The self-balancing two-wheeled robot NXTway-GS (Yamamoto 2015), built on the basis of the LEGO®Mindstorms NXT developer kit, is widely used in education. Also, the telepresence and video conferencing two-wheeled robot Double®(Double 2014) is very popular recently.

The two-wheeled robots have dynamics which is similar to the inverted pendulum dynamics so that they are inherently unstable and should be stabilized around the vertical position using a control system. Linear-quadratic or proportional-integral-derivative (PID) control laws are usually implemented in order to achieve vertical stabilization and desired position in the horizontal plane.

In this section we present the design and experimentation of a robust control system of two-wheeled robot which implements a μ-regulator for vertical stabilization and a proportional-integral controller of the robot rotation around the vertical axis [17]. Due to the lack of accurate analytical robot model, the control system design is done by using a model built with the aid of an identification procedure. A discrete-time Kalman filter of 2nd order is implemented to estimate the plant dynamics around the vertical axis. A software in MATLAB®/Simulink®environment is developed for generation of control code which is embedded in the Texas Instruments Digital Signal Controller

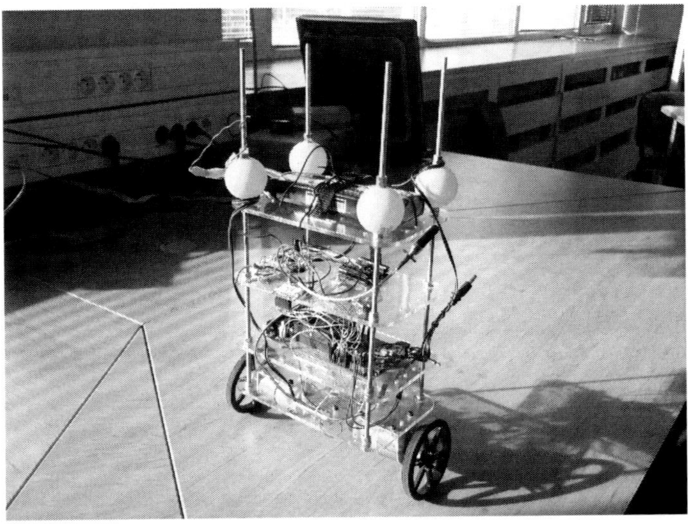

Figure 18. Two-wheeled robot in self-balancing mode.

TMS320F28335. Results from the simulation of the closed-loop system as well as experimental results obtained during the real-time implementation of the controller designed are given.

5.1. Robot Description

The robot is equipped with two servo drives for actuation, MEMS inertial sensor ADIS16350 for measuring the angular velocities $\dot{\phi}$ and $\dot{\psi}$ of robot body in the vertical plane and around the vertical axis, respectively, quadrature encoders for measuring the position of the wheels and a digital signal controller Texas Instruments TMS320F28335 implementing a discrete real-time stabilization algorithm with sampling period $T_0 = 0.005$ s. The robot balancing is achieved by rotating the wheels in appropriate direction. The computation of control actions to both DC brushed drive motors is realized in single precision on the basis of signals from the gyro sensor measuring the angular rate (and, after integration, the tilt angle ϕ) and signals from rotary encoders measuring the wheels rotation angles. The control of the DC motors is executed by Pulse Width Modulated (PWM) signals. A 2nd order Kalman filter is used to estimate the yaw angle ψ.

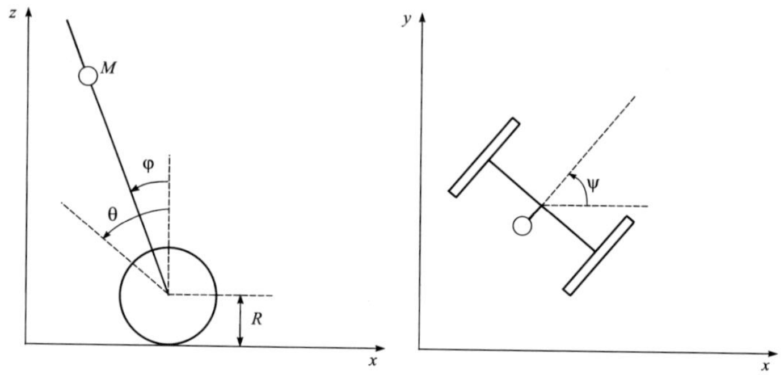

Figure 19. Robot motion in vertical and horizontal planes.

5.2. Model Identification

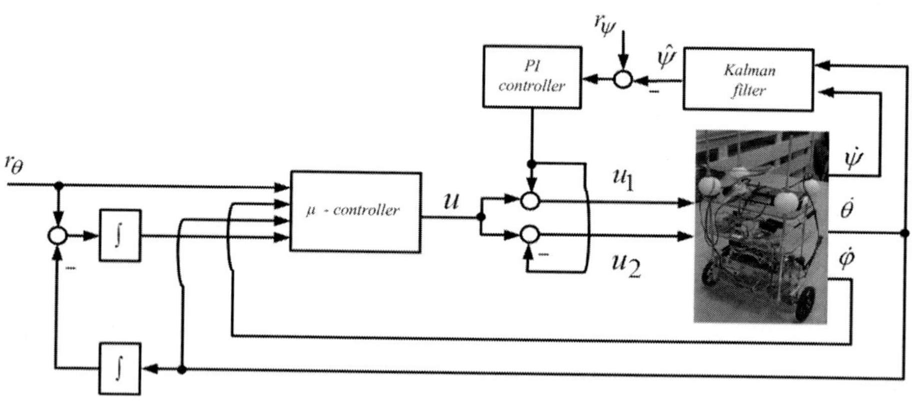

Figure 20. Block-diagram of the control system.

The block-diagram of the two-wheeled robot control system is shown in Figure 20.

In order to simplify the robot dynamics it is assumed that the motions in vertical and horizontal planes are independent. This allows to represent the robot dynamics by a single-input single-output plant as shown in Figure 21.

In respect to the stabilization in upper equilibrium state and to the control of forward-backward motion, the robot is described by Auto Regressive Mov-

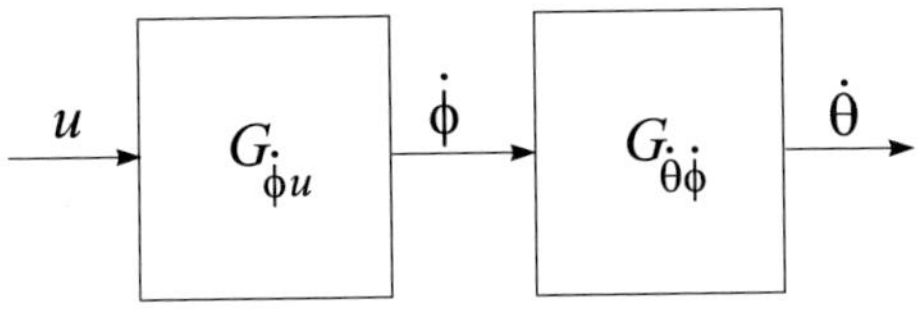

Figure 21. Representation of robot dynamics.

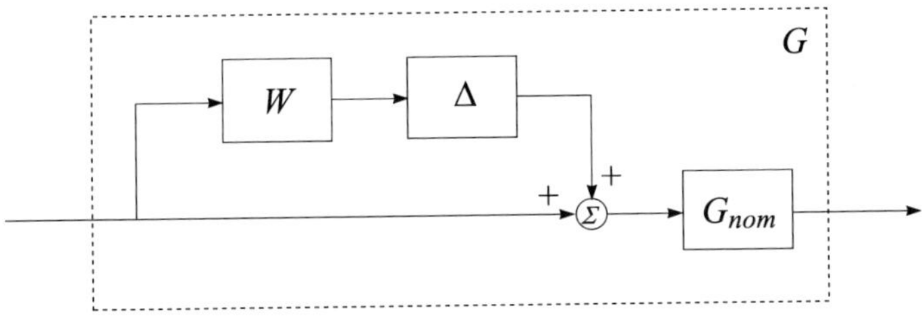

Figure 22. Input multiplicative uncertainty.

ing Average with eXternal input (ARMAX) and Box-Jenkins (BJ) discrete-time models, respectively. For this aim we use the corresponding functions from System Identification Toolbox®3 (Ljung 2013). The ARMAX model of 7th order with structure parameters $na = 7, nb = 7, nc = 7, nk = 3$ describes the link between the control signal u and the rate $\dot{\phi}$, while the BJ model of 3rd order with structure parameters $nb = 3, nf = 3, nc = 3, nd = 3, nk = 1$ gives the link between the rates $\dot{\phi}$ and $\dot{\theta} = (\dot{\theta}_L + \dot{\theta}_R)/2$ where $\dot{\theta}_L$ and $\dot{\theta}_R$ are the angular velocities of left and right wheels, respectively. ARMAX and BJ models, resulting from identification, are assumed as nominal and have the form

$$\dot{\phi}(z) = G_{\dot{\phi}u,nom}(z)u(z) + \nu_{\dot{\phi}}(z), \tag{11}$$

$$\dot{\theta}(z) = G_{\dot{\theta}\dot{\phi},nom}(z) + \nu_{\dot{\theta}}(z) \tag{12}$$

where $G_{\dot{\phi}u,nom}(z)$ is the nominal transfer function from u to $\dot{\phi}$, $G_{\dot{\theta}\dot{\phi},nom}(z)$ is the nominal transfer function from $\dot{\phi}$ to $\dot{\theta}$, the noises $\nu_{\dot{\phi}}, \nu_{\dot{\theta}}$ are obtained during the identification procedure and reflect the uncertainty in the model found. These noises are represented as models with input multiplicative uncertainties which are used in the design of robust controller.

Condition of unbiased parameter estimates guarantees that the exact parameter values are contained in the confidence intervals of estimates with probabil-

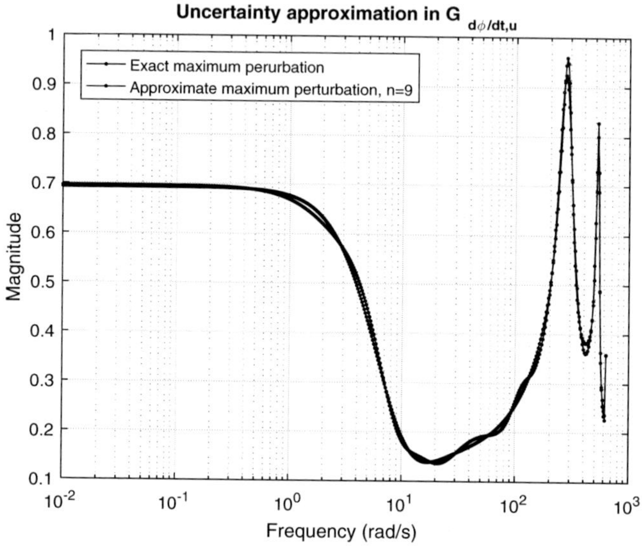

Figure 23. Approximation of the relative uncertainty in $G_{\dot{\phi}u}$.

ity close to 1. This allows derivation of multiplicative uncertainty models of the transfer functions $G_{\dot{\phi}u}$ and $G_{\dot{\theta}\dot{\phi}}$ in the form of LFT shown in Figure 22. Based on confidence intervals, maximum relative deviations from nominal models are obtained. These deviations are approximated, through optimization procedure, with shaping filters which are represented by transfer function of 9th and 5th order, respectively.

The maximum relative uncertainties in $G_{\dot{\phi}u}$ and $G_{\dot{\theta}\dot{\phi}}$ along with their approximations are represented in Figures 23 and 24, respectively.

Resulting uncertain models for the robot vertical and longitudinal motion are

$$G_{\dot{\phi}u}(z) = G_{\dot{\phi}u,nom}(z)(1 + W_{\dot{\phi}u}(z)\Delta_{\dot{\phi}}), \tag{13}$$

$$G_{\dot{\theta}\dot{\phi}}(z) = G_{\dot{\theta}\dot{\phi},nom}(z)(1 + W_{\dot{\theta}\dot{\phi}}(z)\Delta_{\dot{\theta}}) \tag{14}$$

where $W_{\dot{\phi}u}(z), W_{\dot{\theta}\dot{\phi}}(z)$ are the corresponding shaping filters obtained by approximations of the relative magnitude deviations and $\Delta_{\dot{\phi}}$, $\Delta_{\dot{\theta}}$ are uncertainties which satisfy

$$|\Delta_{\dot{\phi}}| < 1, |\Delta_{\dot{\theta}}| < 1.$$

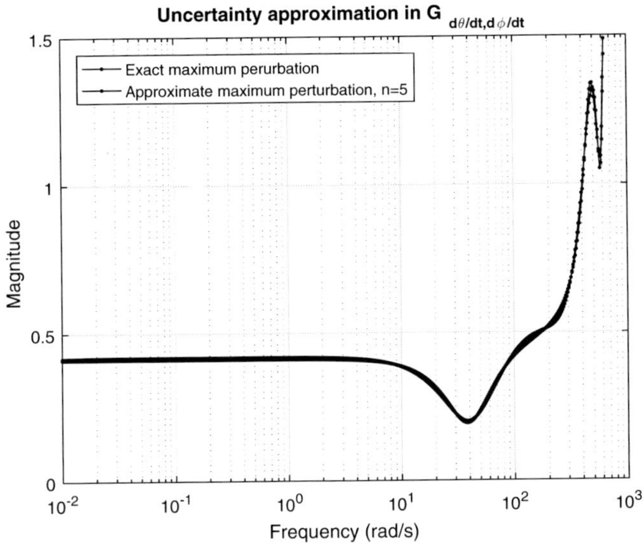

Figure 24. Approximation of the relative uncertainty in $G_{\dot{\theta}\dot{\phi}}$.

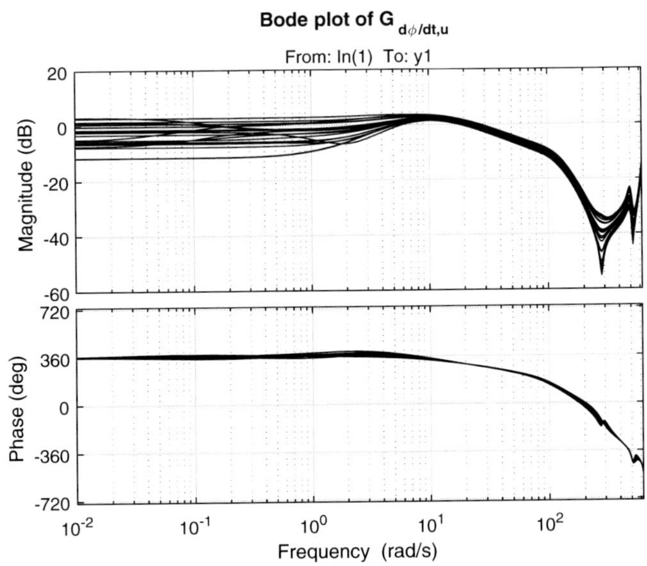

Figure 25. Bode plot of $G_{\dot{\phi}u}$.

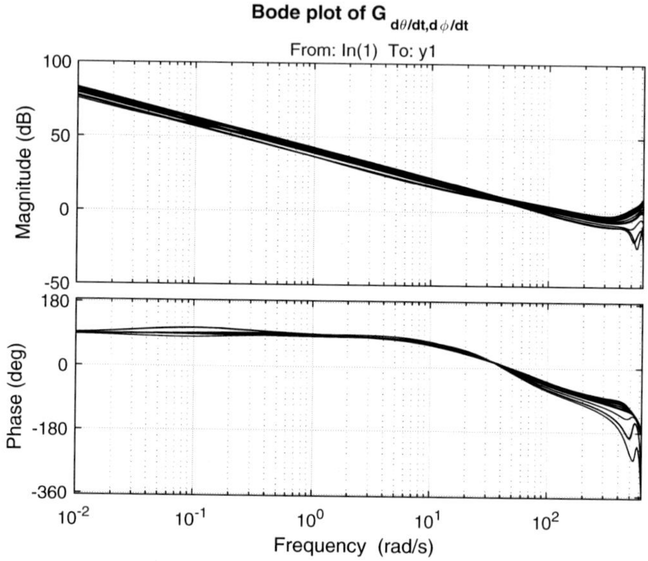

Figure 26. Bode plot of $G_{\dot{\theta}\dot{\phi}}$.

The Bode plots of the uncertain models $G_{\dot{\phi}u}$ and $G_{\dot{\theta}\dot{\phi}}$ are shown in Figures 25 and 26, respectively.

5.3. Controller Design

The closed-loop system block-diagram corresponding to the μ-synthesis problem is shown in Figure 27. In order to obtain better position accuracy a feedback from the integral of the tracking error $r_\theta - \theta$ to the controller is introduced (Goodwin et al. 2001). The control actions to the plant are realized by a Digital Signal Controller in real time with sampling frequency $f_s = 200$ Hz. For this reason the μ-synthesis is implemented to design a discrete-time controller at this sampling frequency.

To obtain good performance of the closed-loop system we shall implement a two-degree-of-freedom controller (Gu et al. 2013). The control actions are generated according to the expression

$$u_c = [K_r \ K_y] \begin{bmatrix} r_\theta \\ y_c \end{bmatrix} = K_r r_\theta + K_y y_c, \tag{15}$$

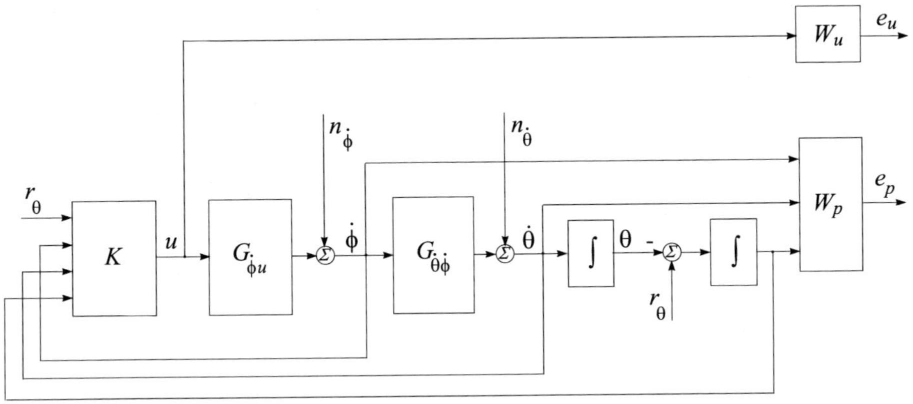

Figure 27. Block-diagram of the closed-loop system.

where r_θ is the wheels reference angle and

$$y_c = \left[\dot{\phi},\ \dot{\theta},\ \int(r_\theta - \theta)\right]^T$$

is the output feedback vector, K_y is the output feedback transfer function matrix and K_r is the pre-filter transfer function.

The weighted closed-loop system outputs e_p and e_u satisfy the equation

$$\begin{bmatrix} e_p \\ e_u \end{bmatrix} = \begin{bmatrix} W_p S_o G K_r \\ W_u S_i K_r \end{bmatrix} r_\theta \tag{16}$$

where the matrix $S_i = (I - K_y G)^{-1}$ is the input sensitivity transfer function matrix, $S_o = (I - G K_y)^{-1}$ is the output sensitivity transfer function matrix and $G = G_{\dot{\theta}\dot{\phi}} G_{\dot{\phi}u}$.

The performance criterion requires the transfer function matrix from the exogenous input signal r_θ to the output signals e_p and e_u to be small in the sense of $\|\cdot\|_\infty$, for all possible uncertain plant models G. This leads to small weighted signals $\dot{\phi}$, $\dot{\theta}$ and $\int(r_\theta - \theta)$ and small control action. The transfer function matrices W_p and W_u are used to reflect the relative importance of the different frequency ranges for which the performance requirements should be fulfilled.

The μ-synthesis is done for several performance weighting functions that ensure a good balance between system performance and robustness. On the basis of the experimental results, we choose the performance weighting function (in the continuous-time case) as

$$W_p(s) = \begin{bmatrix} 52.2\frac{0.005s+1}{s+1} & 0 & 0 \\ 0 & 0.029\frac{0.6s+1}{0.007s+1} & 0 \\ 0 & 0 & 4.7\frac{4s+1}{125s+1} \end{bmatrix}$$

and the control weighting function as

$$W_u(s) = \frac{1}{30000} \frac{\frac{1}{0.007}s+1}{\frac{1}{200}s+1}.$$

The performance weighting functions are chosen as low pass filters, and the control weighting function is chosen as high pass filter with appropriate bandwidth in order to impose constraints on the spectrum of the control actions. These functions are discretized for the sampling frequency of 200 Hz and included in the open-loop system model.

The μ-synthesis is performed by using the MATLAB®function `dksyn` (Balas et al. 2012). In order to obtain better convergence of the D-K iterations two small noises $n_{\dot{\phi}}$ and $n_{\dot{\theta}}$ with intensity 10^{-4} are added to the angular velocities $\dot{\phi}$ and $\dot{\theta}$, respectively. Three iterations are performed that decrease the maximum value of μ to 0.693. The final controller obtained is of 50th order.

The Bode plot of the closed-loop system is shown in Figure 28. It is seen that the closed-loop bandwidth is approximately 1 rad/s.

The magnitude plot of the closed-loop in respect to the tracking error $e_\theta = r_\theta - \theta$ is shown in Figure 29.

In Figures 30 and 31 we show the influence of the noises in $\dot{\phi}$ and $\dot{\theta}$, respectively, to the control action u. It is seen that the effect of noise in $\dot{\phi}$ is much stronger than the effect of noise in $\dot{\theta}$.

Finally, in Figure 32 we show the plots of the structured singular values corresponding to the robust stability and robust performance of the closed-loop system. The system achieves both robust stability and robust performance in respect to the uncertainties corresponding to the identification of robot model.

A PI controller of the angular motion around the vertical axis is also designed. The yaw angular velocity $\dot{\psi}$ is measured by a gyroscope of the same type as the gyro used to measure the tilth rate. This gyroscope contains a significant noise ψ_g which is modeled by the additional equation

$$\dot{\psi}_g(k+1) = \dot{\psi}_g(k) + J_g \nu_g(k) \qquad (17)$$

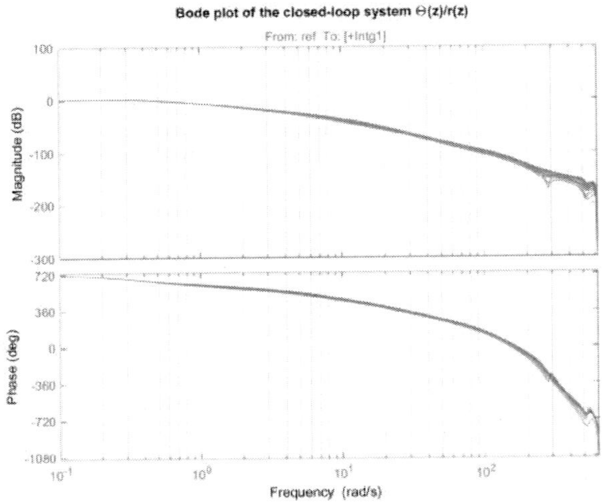

Figure 28. Bode plot of the closed-loop system.

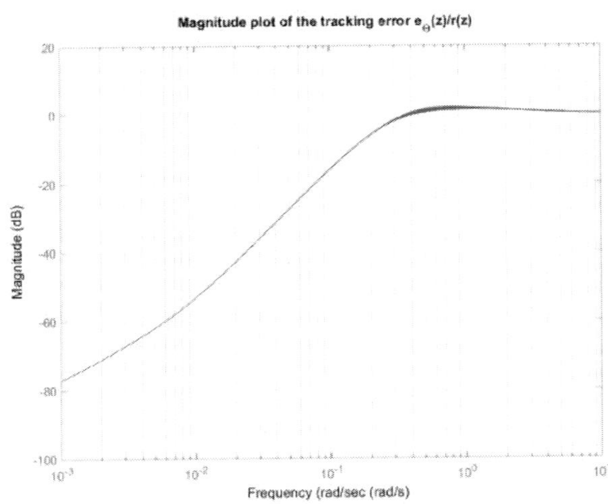

Figure 29. Magnitude plot of the tracking error.

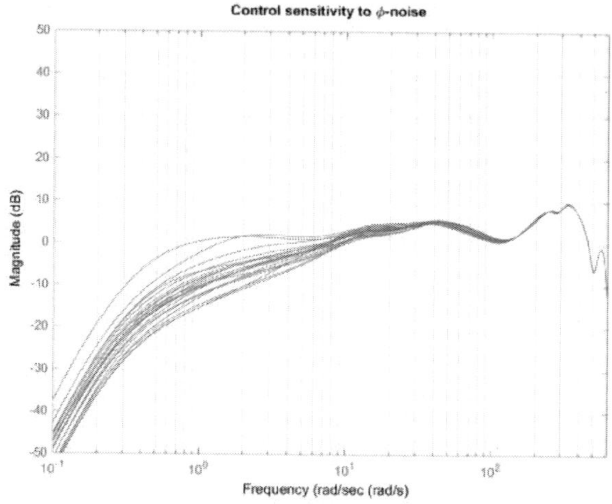

Figure 30. Sensitivity of control to noise in $\dot{\phi}$.

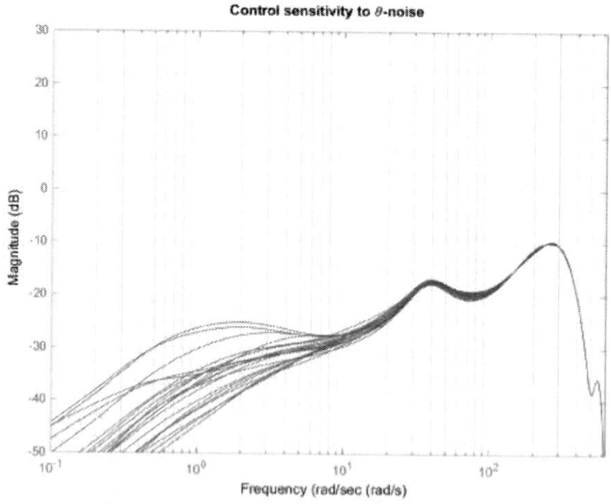

Figure 31. Sensitivity of control to noise in $\dot{\theta}$.

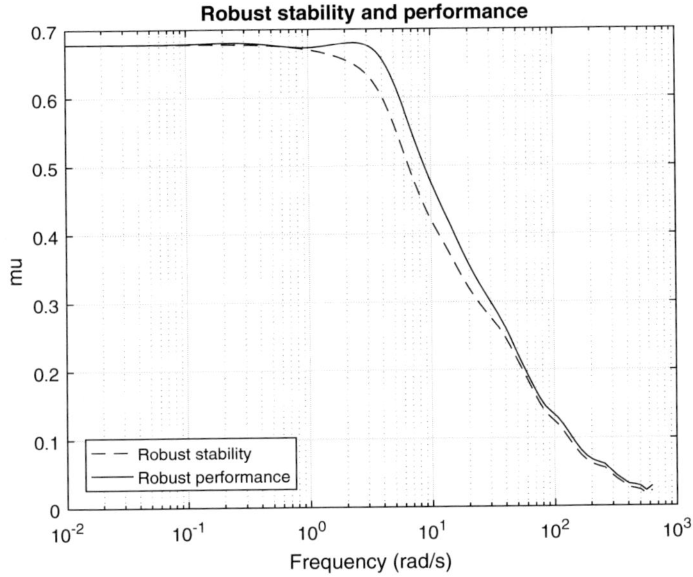

Figure 32. Robust stability and performance of the closed-loop system.

where ν_g is a white gaussian noise with unit variance and the coefficient J_g is determined experimentally to obtain a good estimate of $\dot{\psi}$ as $J_g = 0.001$. A second order Kalman filter is designed to produce sufficiently accurate estimate $\hat{\psi}$ of the yaw angle.

5.4. Experimental Results

A simulation scheme of the control system and a specialized software in MATLAB®/Simulink®environment is developed to implement the control code. To simplify the controller, its order is reduced from 50 to 30. With the aid of Simulink Coder®[28] and Code Composer Studio®, a code is generated from this software which is embedded in the Texas Instruments Digital Signal Controller TMS320F28335 [27].

Several experiments with the controller designed are performed and comparison with the simulation results is done. The experimental results obtained during the real-time robot control are given in Figures 33 - 36.

It is seen form the Figures33 and 34 that the wheels track accurately the

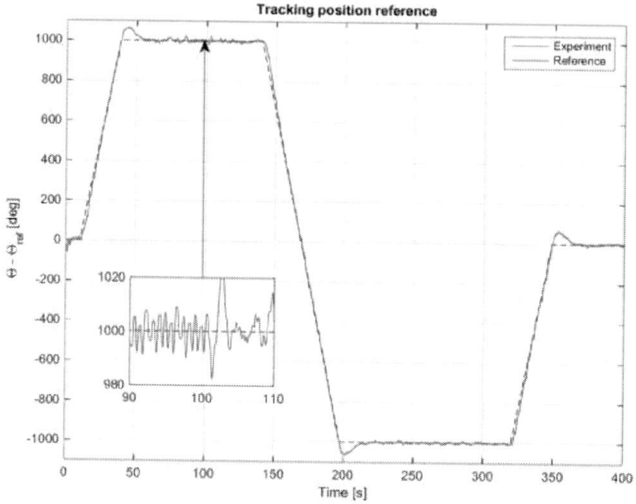

Figure 33. Tracking position reference.

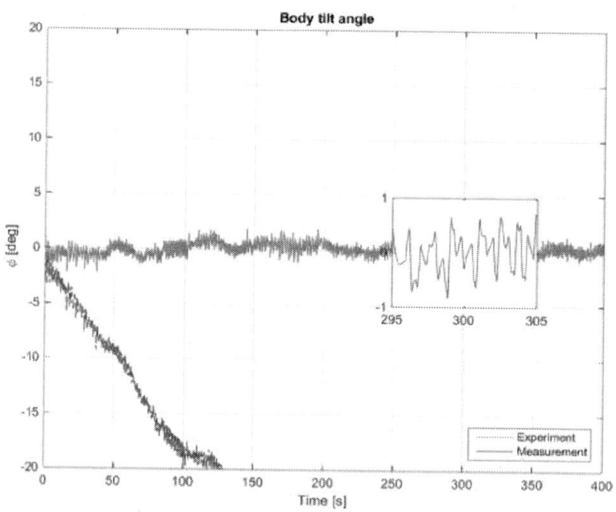

Figure 34. Body angle variation.

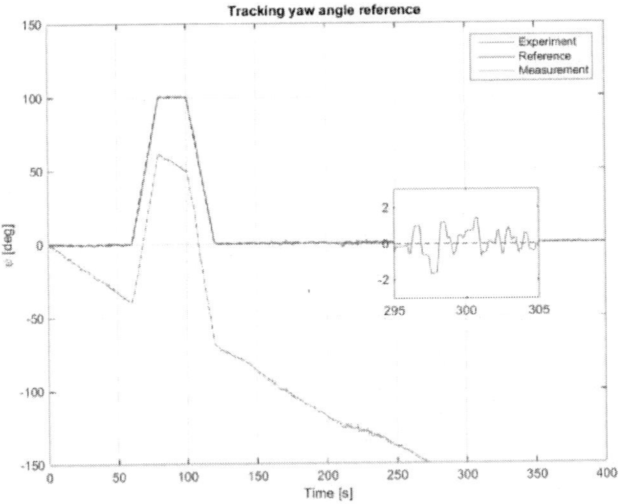

Figure 35. Tracking yaw angle reference.

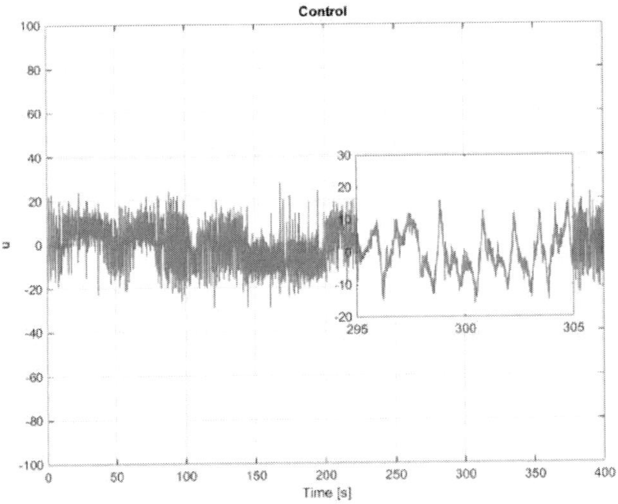

Figure 36. Control action.

reference and the robots keeps well its vertical position. The measured angle ϕ contains a significant bias which is removed by using off-line a 17th order Kalman filter. It is interesting to note that this bias does not affect the vertical stabilization of the robot which is due to the good filtration properties of the system with μ-controller. The usage of the estimate $\hat{\psi}(k)$, obtained by a 2nd order Kalman filter, instead of $\psi(k)$ ensures exact rotation of the robot around the vertical axis, while there is a increasing with the time bias in the measured value of $\psi(k)$ due to the integration of the gyro noise (Figure 35). The control action does not exceed 40 units with maximum allowable value equal to 50 (Figure 36).

Conclusion

The paper presents a brief survey of several issues arising in the design and implementation of robust control laws in embedded control systems. Some difficulties and unsolved problems that require further research are pointed out. A case study of robot control design is presented which illustrates the progress in the implementation of high order robust controllers. Several references are included which may help the reader to find more detailed information about the topics discussed.

References

[1] Allan, D.W. (1966). Statistics of atomic frequency standards. *Proceedings of the IEEE* 54, 221-230. DOI: 10.1109/PROC.1966.4634.

[2] Balas, G., Chiang, R., Packard, A., Safonov, M. (2013). Robust Control Toolbox® 3 , User's Guide, The MathWorks, Inc., Natick, MA. Available from http://www.mathworks.com/help/pdf_doc/robust/robust_ug.pdf.

[3] Bautista-Quintero, R., Pont, M.J. (2008). Implementation of H-infinity control algorithms for sensor-constrained mechatronic systems using low-cost microcontrollers. *IEEE Transactions on Industrial Informatics* 4, 175-184. DOI: 10.1109/TII.2008.2002703.

[4] Bittanti, S., Garatti, S. (2012). System Identification and Control: a fruitful cooperation over half a century, and more. *Proceedings of the 31st Chinese*

Control Conference, July 25-27, 2012, Hefei, China. IEEE, 6-15. ISBN: 978-1-4673-2581-3.

[5] Campion, G., Chung, W. (2008). Wheeled Robots. In *B. Siciliano and O. Khatib, editors, Springer Handbook of Robotics*, chapter 17, pp. 391-410, Springer, Berlin. ISBN: 978-3-540-23957-4, DOI: 10.1007/978-3-540-30301-5.

[6] Chan, R.P.M., Stol, K.A., Halkyard, C.R. (2013). Review of modelling and control of two-wheeled robots. *Annual Reviews in Control* 37, 89-103. DOI: 10.1016/j.arcontrol.2013.03.004.

[7] Double Robotics (2014). Sunnyvale, CA. http://www.doublerobotics.com.

[8] El-Sheimy, N., Hou, H., Niu, X. (2008). Analysis and modeling of inertial sensors using Allan variance. *IEEE Transactions on Instrumentation and Measurement* 57, 140-149. DOI: 10.1109/TIM.2007.908635.

[9] Gahinet, P., Nemirovski, A., Laub, A.J., Chilali, M. (1995). *LMI Control Toolbox*, The MathWorks Inc., Natick, MA.

[10] Goodwin, G.C., Graebe, S.F., Salgado, M.E. (2001). *Control System Design*, Prentice Hall, Upper Saddle River, NJ. ISBN: 9780139586538.

[11] Gu, D.-W., Petkov, P.Hr., Konstantinov, M.M. (2013). *Robust Control Design with MATLAB®, 2nd ed.*. Springer-Verlag, London. ISBN: 978-1-4471-4681-0.

[12] Gugercin, S., Antoulas, A. C., Zhang, H. P. (2003). An approach to identification for robust control. *IEEE Transactions on Automatic Control* 48, 1109-1115. DOI: 10.1109/TAC.2003.812821.

[13] Howlader, A.M., Urasaki, N., Yona, A., Senjyu, T., Saber, A.Y. (2013). Design and implement a digital H_∞ robust controller for a MW-class PMSG-based grid-interactive wind energy conversion system. *Energies* 6, 2084-2109. DOI:10.3390/en6042084.

[14] Hristu-Varsakelis, D., Levine, W.S. (Eds.) (2005). *Handbook of Networked and Embedded Control Systems*. Birkhäuser, Boston. ISBN: 978-0-8176-3239-7.

[15] Isermann, R., Münchhof, M. (2011). *Identification of Dynamic Systems. An Introduction with Applications*. Springer-Verlag, Berlin. ISBN: 978-3-540-78878-2, DOI: 10.1007/978-3-540-78879-9.

[16] Jahaya, J., Nawai, S.W., Ibrahim, Z. (2011). Multi input single output closed loop identification of two wheel inverted pendulum mobile robot. *2011 IEEE Student Conference on Research and Development (SCOReD)*, 19-20 Dec. 2011, Cyberjaya, Malaysia. IEEE, 138-143. DOI: 10.1109/SCOReD.2011.6148723.

[17] Kralev, J., Slavov, Ts., Petkov, P. (2016). Design and experimental evaluation of robust controllers for a two-wheeled robot. *International Journal of Control* 89. DOI: 10.1080/00207179.2016.1151940.

[18] Landau, I.D., Zito, G. (2006). *Digital Control Systems. Design, Identification and Implementation*. Springer-Verlag, London. ISBN 978-1-84628-055-9.

[19] Ljung, L. (1999). *System Identification: Theory for the User, 2nd ed.*. Prentice-Hall, Inc. Englewood Cliffs, NJ. ISBN: 978-0136566953.

[20] Ljung, L. (2013) *System Identification Toolbox® User's Guide*. The MathWorks Inc., Natick, MA. ISBN 0-13-881640-9.

[21] Lunze, J., Lamnabhi-Lagarrigue, F. (Eds.) (2009). *Handbook of Hybrid Systems Control. Theory, Tools, Applications*. Cambridge University Press, Cambridge. ISBN: 978-0-521-76505-3.

[22] Lupián L.F., Avila, R. (2009). Stabilization of a wheeled inverted pendulum by a continuous-time infinite-horizon LQG optimal controller. *IEEE Latin American Robotic Symposium LARS'2008*, Natal, Rio Grande do Norte, Brazil, 29-30 Oct. 2008. IEEE, 65-70. DOI: 10.1109/LARS.2008.33.

[23] Muhammad, M., Buyamin, S., Ahmad, M.N., Nawawi, S.W. (2011). Dynamic modeling and analysis of a two-wheeled inverted pendulum robot. *Third International Conference on Computational Intelligence, Modelling and Simulation CIMSiM 2011*, 20-22 Sept. 2011, Langkawi, Malaysia. IEEE, 159-164. DOI: 10.1109/CIMSim.2011.36.

[24] Nawawi, Ahmad, Osman (2010). Real-time control system for a two-wheeled inverted pendulum mobile robot. Chapter 6 in: I. Fürstner (Ed.), *Advanced Knowledge Application in Practice*, Sciyo, Rijeka, Croatia, 299-312. ISBN 978-953-307-141-1.

[25] Quinchia, A.G., Ferrer, C., Falco, G., Falletti, E., Dovis, F. (2012). Analysis and modelling of MEMS Inertial Measurement Unit. *2012 International Conference on Localization and GNSS (ICL-GNSS)*, 25-27 June 2012, Starnberg, Germany. IEEE, 1-7. DOI: 10.1109/ICL-GNSS.2012.6253129.

[26] The MathWorks, Inc. (2013). *Embedded Coder*. Natick, MA. Available from: http://www.mathworks.com/products/datasheets/pdf/embedded-coder.pdf.

[27] Spectrum Digital, Inc. (2007). *eZdspTMF28335 Technical Reference*. Available from: http://c2000.spectrumdigital.com/ezf28335/docs/ezdspf28335c_techref.pdf.

[28] The MathWorks, Inc. (2013). *Simulink Coder*. Natick, MA. Available from: http://www.mathworks.com/products/datasheets/pdf/simulink-coder.pdf.

[29] IEEEStd 952-1997 (1998). *IEEE Standard Specification Format Guide and Test Procedure for Single-Axis Interferometric Fiber Optic Gyros*. IEEE Aerospace and Electronic Systems Society.

[30] Raafat, S.M., Akmeliawati, Abdulljabar, I. (2012). Robust H_∞ controller for high precision positioning system, design, analysis, and implementation. *Intelligent Control and Automation* 3, 262-273. DOI:10.4236/ica.2012.33030.

[31] Sánchez-Peña, R.S., Sznaier, M. (1998). *Robust Systems. Theory and Applications*, John Wiley & Sons, Inc., New York. ISBN: 978-0-471-17627-5.

[32] Segway Personal Transporters (2012). Bedford, NH. http://www.segway.com.

[33] Skogestad, S., Postlethwaite, I. (2005). *Multivariable Feedback Control, 2nd ed.*. John Wiley and Sons Ltd, Chichester, UK. ISBN: 978-0-470-01167-6.

[34] Van den Hof, P. (2001). Identification of experimental models for control design. *Proceedings of the 18th Instrumentation and Measurement Technology Conference, (IMTC 2001)*, 21-23 May 2001, Budapest, Hungary. IEEE, vol. 2, 1155 - 1162. DOI: 10.1109/IMTC.2001.928260.

[35] Venkatesh, S. (2003). Identification of uncertain systems described by Linear Fractional Transformations. *Proceedings of the 42nd IEEE Conference an Decision and Control*, Maui, Hawaii USA, December 2003. IEEE, 5532-5537. DOI: 10.1109/CDC.2003.1272518.

[36] Yamamoto, Y. (2013). NXTway-GS (Self-Balanced Two-Wheeled Robot Controller Design).
http://www.mathworks.com/matlabcentral/fileexchange/.

[37] Zhou, K., Doyle, J.C., Glover, K. (1996). *Robust and Optimal Control*. Prentice Hall, Upper Saddle River, NJ. ISBN: 0-13-456567-3.

In: Robust Control: Systems, Theory and Analysis ISBN: 978-1-53610-826-2
Editor: Sharon Bennett

Chapter 2

PARETO SUBOPTIMAL SOLUTIONS IN MULTI-OBJECTIVE CONTROL AND FILTERING PROBLEMS

Dmitry V. Balandin * ***and Mark M. Kogan*** †
Lobachevsky Nizhny Novgorod State University, Russia,
Architecture and Civil Engineering,
State University of Nizhny Novgorod, Russia

Abstract

A multi-objective disturbance attenuation problem is considered as a framework for control and filtering problems under multiple exogenous disturbances and unknown initial conditions. Necessary conditions of the Pareto optimality in this problem are derived. Pareto suboptimal solutions are shown to be optimal solutions with respect to a multi-objective cost which is a parameterized disturbance attenuation level of the system under disturbances acting on all channels simultaneously. It is established that the relative losses of these Pareto suboptimal solutions compared to the Pareto optimal ones do not exceed $1 - \sqrt{N}/N$, where N is the number of criteria. This approach is applied to control and filtering problems in continuous-time and discrete-time systems with multiple classes of L_2-bounded and impulsive disturbances for which the disturbance attenuation levels are H_∞ and γ_0 norms. The H_∞/γ_0 optimal controllers

* E-mail address: dbalandin@yandex.ru
† Email address: mkogan@nngasu.ru

and filters as the Pareto suboptimal solutions are synthesized in terms of linear matrix inequalities (LMIs). All these results are extended to the case when the disturbances acting on different inputs are combined into coalitions. Moreover, it is shown that multi-objective problems for systems with a single disturbance input and N objective outputs under the H_∞ and generalized H_2 norms as criteria are reduced to the case considered. Illustrative examples demonstrate the effectiveness of the approach proposed.

Keywords: multi-objective control and filtering, Pareto optimal solutions, coalition of disturbances, H_∞ norm, γ_0 norm, H_∞/γ_0 optimal control, LMIs

1. Introduction

This chapter deals with multi-objective disturbance attenuation control and filtering problems for disturbances from different classes. We consider linear systems with multiple disturbance inputs and a single objective output. A disturbance attenuation level which is the worst-case ratio of the objective output norm and a disturbance norm from the corresponding class is defined for each channel. The problem to be solved is to find a feedback controller or a filter minimizing these disturbance attenuation levels in sense of Pareto optimality.

The multi-objective tradeoff approach is very useful in engineering practice when there are some competing objectives. It is well known that multi-objective problems traditionally are extremely difficult to solve and find the Pareto optimal set which is defined as a set of unimprovable alternatives. The most basic multi-objective techniques involve the scalarization of the vector of objectives using weighted sums. A set of weights is chosen to transform the problem into a standard optimisation problem with respect to some multi-objective cost. By adjusting the weights, a set of solutions can be generated. When the objectives are convex, very determinative conclusions can be obtained as to feasibility of certain combinations of cost functions.

Despite a huge amount of publications in the field of multi-objective control problems with H_2 and H_∞ criteria there are no results concerning the exact Pareto optimal set and a comparison of various approximate sets with the exact one. The exact solution of this problem is understood as minimizing the H_2 performance measure subject to constraint imposed on the H_∞ performance measure. Instead of minimizing the H_2 norm of the transfer function for one channel, an 'upper bound' for the H_2 norm is considered subject to the H_∞-

constraint on the transfer function for the other channel. This approach was suggested for the first time in [1], where the H_∞ performance constraint is embedded within the optimization process by replacing the covariance Lyapunov equation by a Riccati equation whose solution leads to an upper bound on the H_2 performance. The mixed H_2/H_∞ problem includes two different linear matrix inequality (LMI) constraints that make the problem non-convex. In order to recover convexity a common Lyapunov function is used by equating the Lyapunov matrices for the H_2 and H_∞ performances [2, 3, 4, 5, 6]. Of course, this leads to a conservative result. This conservatism was demonstrated by [7, 8] with using genetic algorithms.

The paper [9] was one of the first in which the multi-objective problem for LQG objectives has been considered and the observer-based Pareto optimal solution has been obtained. The authors of the paper [10] have studied the multi-objective H_2 problem using Youla parametrization. They have presented the Pareto optimal controller in the terms of an infinite dimensional parameter Q. In special cases of a single input and multiple outputs or multiple inputs and a single output, the Riccati-based Pareto optimal solutions in the multi-objective H_2 problem were derived. In the paper [11], the above approach based on Youla parametrization has been extended to multi-objective problems with H_2 and H_∞ criteria and a finite dimensional approximation for the Pareto optimal controllers has been proposed. For synthesizing these controllers the authors of the paper [11] have used LMI characterizations for H_2 and H_∞ norms and the ideas of the paper [12] for solving the mixed H_2/H_∞ problem.

The main contribution of this chapter is a formulation and a proof of a necessary condition for Pareto optimality, synthesizing approximate Pareto optimal solutions and estimating a difference between exact and approximate solutions. Specifically, we introduce two performance measures, the optimal and suboptimal multi-objective costs. The minimization of the first one allows to cover the Pareto optimal solutions but is difficult to calculate, while the minimization of the second one can be performed by standard LMI procedures and allows to approximate the Pareto optimal set. The relative losses of the Pareto suboptimal solutions with respect to the Pareto optimal ones based on a comparison of the above costs turn out to be less than or equal to $1-\sqrt{N}/N$, where N is a number of criteria. This concept is extended to the case when the disturbances acting on different inputs are combined into coalitions.

This approach is applied to multiple classes of L_2-bounded and impulsive disturbances in continuous-time and discrete-time systems with H_∞ and γ_0 norms as the disturbance attenuation levels. These classes are very important in the problems of optimal protection of the objects from shock and vibration (see, for example, [13, 14]). The γ_0 norm was utilized in papers [15, 16] to characterize the worst-case ratio of the objective output norm and Euclidian norm of an intensity of impulsive disturbance. For these criteria the suboptimal multi-objective cost turns out to be the H_∞/γ_0 norm that is very closed to the mixed H_2/H_∞ norm. In the simplest situation, the H_∞/γ_0 norm is the worst-case output norm in response to two input disturbances in different channels, one of which is a L_2-bounded signal and the other is an impulsive signal. Such a problem would also naturally arise in considering tradeoff between nominal performance and robustness. It has been shown in [17] that this norm also characterizes the disturbance attenuation level in discrete-time systems under deterministic signals with a bounded energy and stationary Gaussian white zero-mean signals with unknown covariances, acting on two different inputs. The H_∞/γ_0 norm is a generalization of so called H_∞ norm with transients that was introduced and studied in the papers [18, 19, 20, 21]. In this chapter, we show how to synthesize the Pareto suboptimal controllers and filters on the basis of the H_∞/γ_0 norm in terms of LMIs and compute their relative losses compared to true Pareto optimal solutions. The relationship between the H_∞/γ_0 and mixed H_2/H_∞ norms is also revealed. It is shown by duality that the multi-objective problems for systems with a single disturbance input and multiple objective outputs for which criteria are the H_∞ and/or generalized H_2 norms can be reduced to the multi-objective problems considered.

The chapter is organized as follows. Section 1 is an introduction. Section 2 describes criteria in the form of disturbance attenuation levels to be considered. Statements of the multi-objective control and filtering problems are in Section 3. Section 4 contains the formulation and the proof of the necessary condition for Pareto optimality in the general multi-objective disturbance attenuation problem as well as estimating the relative losses of the Pareto suboptimal solutions. Section 5 deals with the case of possible coalitions of disturbances. Section 6 and 7 are devoted to synthesizing Pareto suboptimal controllers and filters in the multi-objective problem with H_∞ and γ_0 criteria for continuous-time and discrete-time systems, respectively. The dual multi-objective control problem for systems with a single disturbance input and multiple objective outputs is discussed in Section 8. Section 9 provides with some illustrative examples and Section 10 concludes the chapter.

2. Preliminary: Disturbance Attenuation Levels

Consider an internally stable system described by the equations

$$\begin{aligned} \partial x &= Ax + Bv \\ z &= Cx + Dv, \end{aligned} \tag{2.1}$$

where ∂ is the differentiation/delay operator for continuous-time/discrete-time systems, $x \in R^{n_x}$ is the state, $v \in R^{n_v}$ is the disturbance, $z \in R^{n_z}$ is the objective output. Let H be a transfer matrix of this system from v to z.

If the disturbance is a signal with the bounded $L_2(l_2)$-norm and the 'size' of the output is the $L_2(l_2)$-norm as well, it is known that

$$\begin{aligned} \sup_{v \in L_2} \frac{\|z\|}{\|v\|} &= \sup_{\omega \in [0,\infty)} \lambda_{max}^{1/2}[H^T(-i\omega)H(i\omega)] = \|H\|_\infty, \\ \sup_{v \in l_2} \frac{\|z\|}{\|v\|} &= \sup_{\varphi \in [0,2\pi)} \lambda_{max}^{1/2}[H^T(e^{-i\varphi})H(e^{i\varphi})] = \|H\|_\infty \end{aligned} \tag{2.2}$$

for the continuous-time and discrete-time systems, respectively.

Recall that the H_2-norm of this system with $D = 0$ is defined as follows

$$\|H\|_2^2 = \text{tr}\,(S_c(H)) = \text{tr}\,(B^T X_c B), \quad \|H\|_2^2 = \text{tr}\,(S_d(H)) = \text{tr}\,(B^T X_d B), \tag{2.3}$$

where

$$S_c(H) = \frac{1}{2\pi} \int_{-\infty}^{\infty} H^T(-i\omega)H(i\omega)\,d\omega = \int_0^{\infty} B^T e^{A^T t} C^T C e^{At} B\,dt,$$

$$S_d(H) = \frac{1}{2\pi} \int_0^{2\pi} H^T(e^{-i\varphi})H(e^{i\varphi})\,d\varphi = \sum_{t=0}^{\infty} B^T (A^T)^t C^T C A^t B,$$

each of the matrices $X_c = X_c^T > 0$ and $X_d = X_d^T > 0$ is a solution of the corresponding Lyapunov equation

$$A^T X + XA + C^T C = 0, \quad A^T X A - X + C^T C = 0. \tag{2.4}$$

If the disturbance is a signal with the bounded $L_2(l_2)$-norm and the 'size' of the output is the $L_\infty(l_\infty)$-norm, i.e. $\|z\|_\infty = \sup_{t \geq 0} |z(t)|$, then the disturbance attenuation level is defined as follows

$$\sup_{v \in L_2(l_2)} \frac{\|z\|_\infty}{\|v\|} = \|H\|_{g2}.$$

This is a variant of the generalized H_2 norm introduced by Wilson [22] (see also [23, 24, 25]). It is computed as follows

$$\|H\|_{g2}^2 = \lambda_{max}(CY_cC^T) = \lambda_{max}[S_c(H^*)], \quad \|H\|_{g2}^2 = \lambda_{max}(CY_dC^T) = \lambda_{max}[S_d(H^*)], \tag{2.5}$$

where H^* is the complex conjugate matrix to matrix H, each of the matrices $Y_c = Y_c^T > 0$ and $Y_d = Y_d^T > 0$ is a solution of the corresponding Lyapunov equation

$$AY + YA^T + BB^T = 0, \quad AYA^T - Y + BB^T = 0. \tag{2.6}$$

If the disturbance is an impulsive signal of the form $v = d\delta(t)$ in the continuous-time case, where $\delta(t)$ is the Dirac delta function, and $v(t) = d\delta_d(t)$, $\delta_d(0) = 1$ and $\delta_d(t) = 0$ for $t \geq 1$ in the discrete-time case, then the disturbance attenuation level is defined as follows

$$\sup_{d \neq 0} \frac{\|z\|}{|d|} = \gamma_0(H). \tag{2.7}$$

This is an induced norm of the linear operator generated by the system, which maps vectors d into outputs z. Since $z(t) = Ce^{At}Bd$ or $z_t = CA^{t-1}Bd$ for $t = 1, \ldots$, it is easy to see that

$$\gamma_0^2(H) = \lambda_{max}(B^TX_cB) = \lambda_{max}[S_c(H)], \quad \gamma_0^2(H) = \lambda_{max}(B^TX_dB) = \lambda_{max}[S_d(H)], \tag{2.8}$$

where each of the matrices $X_c = X_c^T > 0$ and $X_d = X_d^T > 0$ is the solution of the corresponding Lyapunov equation in (2.4). The formulas (2.3) and (2.8) imply the following relations between the H_2 and γ_0 norms:

$$\frac{1}{\sqrt{n_v}}\|H\|_2 \leq \gamma_0 \leq \|H\|_2,$$

where n_v is the dimension of the disturbance input v. If $n_v = 1$, then $\gamma_0 = \|H\|_2$. It is important to emphasize that the H_2 norm is the objective output norm in response for impulsive disturbances applied one at a time at each disturbance channel, while the γ_0 norm corresponds to the worst-case objective output norm over the possible disturbance directions.

Consider the dual system

$$\begin{aligned} \partial x &= A^Tx + C^Tv \\ z &= B^Tx + D^Tv \end{aligned} \tag{2.9}$$

with the transfer matrix $\hat{H} = H^*$ from the input v to the output z. In view of (2.5) and (2.8) we get the generalized H_2 norm of the system (2.1) coincides with the γ_0 norm of the dual system (2.9), i.e. $\|H\|_{g2} = \gamma_0(H^T)$. In addition, note that from (2.2) and (2.3) it immediately follows that both the H_∞ norms as well as the H_2 norms of the systems (2.1) and (2.9) coincide.

3. Multi-Objective Control and Filtering Problems with H_∞ and/or γ_0 Criteria

We present the multi-objective filtering and control problems in the unified framework as follows. Consider an internally stable linear time-invariant system parameterized by the parameter matrix Θ and described by the equations

$$\begin{aligned} \partial x &= A(\Theta)x + \sum_{i=1}^{m} B_i(\Theta)w_i + \sum_{j=m+1}^{N} B_j(\Theta)w_j, \quad x(0) = 0, \\ z &= C(\Theta)x + \sum_{i=1}^{m} D_i(\Theta)w_i, \end{aligned} \tag{3.1}$$

where $x \in R^{n_x}$ is the state, $z \in R^{n_z}$ is the objective output. Let $H_k(\Theta)$ be a transfer matrix of the system from the disturbance input $w_k \in R^{n_k}$ to z, $k = 1, \dots, N$. The criteria to be minimized are as follows

$$J_i(\Theta) = \|H_i(\Theta)\|_\infty,\ i = 1, \dots, m, \quad J_j(\Theta) = \gamma_0(H_j(\Theta)),\ j = m+1, \dots, N.$$

A key concept in multi-objective optimization is the Pareto optimal set. The set $\mathcal{P} = \{\Theta_P\}$ is the Pareto optimal if there is not a matrix Θ such that the inequalities $J_k(\Theta) \le J_k(\Theta_P)$, $k = 1, \dots, N$, with at least one of the inequalities being strict, be valid. The problem is to characterize Pareto optimal solutions, i.e.

$$\Theta_P = \arg\min_{\Theta}\{\|H_i(\Theta)\|_\infty,\ \forall i;\ \gamma_0(H_j(\Theta)),\ \forall j\}. \tag{3.2}$$

Such a problem can be regarded as a game between the nature which selects a signal from one of the specified classes to direct it at the corresponding channel and a person who chooses the matrix Θ in such a way to minimize his losses. Note that the case when there is only a single type of the criteria in (3.2) is allowed as well.

A traditional way in the multi-objective optimization to solve this problem would be that of minimizing the multi-objective cost

$$J(\Theta) = \sum_{i=1}^{m} \lambda_i \gamma_i^2 + \sum_{j=m+1}^{N} \lambda_j \gamma_j^2, \quad \lambda_k > 0, \quad \sum_{k=1}^{N} \lambda_k = 1$$

subject to constraints $J_i(\Theta) < \gamma_i^2$, $i = 1, \ldots, m$, $J_j(\Theta) < \gamma_j^2$, $j = m + 1, \ldots, N$. Since $\|H_i(\Theta)\|_\infty < \gamma_i$ if and only if there exists $X_i = X_i^T > 0$ such that

$$\begin{pmatrix} A^T(\Theta)X_i + X_i A(\Theta) & \star & \star \\ B_i^T(\Theta)X_i & -\gamma_i^2 I & \star \\ C(\Theta) & D_i(\Theta) & -I \end{pmatrix} < 0$$

for continuous-time systems or

$$\begin{pmatrix} -X_i & \star & \star & \star \\ A^T(\Theta)X_i & -X_i & \star & \star \\ B_i^T(\Theta)X_i & 0 & -\gamma_i^2 I & \star \\ 0 & C(\Theta) & D_i(\Theta) & -I \end{pmatrix} < 0$$

for discrete-time systems and $\gamma_0(H_j(\Theta)) < \gamma_j$ if and only if there exists $X_j = X_j^T > 0$ such that

$$\begin{pmatrix} A^T(\Theta)X_j + X_j A(\Theta) & \star \\ C(\Theta) & -I \end{pmatrix} < 0, \quad \begin{pmatrix} -X_j & \star \\ B_j(\Theta)X_j & -\gamma_j^2 I \end{pmatrix} < 0$$

for continuous-time systems or

$$\begin{pmatrix} -X_j & \star & \star \\ A^T(\Theta)X_j & -X_j & \star \\ 0 & C(\Theta) & -I \end{pmatrix} < 0, \quad \begin{pmatrix} -X_j & \star \\ B_j(\Theta)X_j & -\gamma_j^2 I \end{pmatrix} < 0$$

for discrete-time systems, these constraints involve product terms between Lyapunov variables X_k and matrix Θ and, therefore, become, in general, nonconvex. In order to recover convexity a common Lyapunov function is used by equating the Lyapunov matrices for different performances, which leads to a conservative result. There are as yet no results on the degree of this conservatism. The approach proposed here allows to search for the Pareto suboptimal solutions and evaluate their relative losses compared to the Pareto optimal solutions.

4. Pareto Optimal and Suboptimal Multi-Objective Costs

Consider a linear time-invariant internally stable dynamic system parameterized by the parameter matrix Θ, in which $\xi_k \in R^{n_k}$, $k = 1, \ldots, N$ are the exogenous disturbance inputs, and $z \in R^{n_z}$ is the objective output. Associated to each channel from the input ξ_k to the output z is defined the disturbance attenuation level

$$J_k(\Theta) = \sup_{\|\xi_k\|_{\Xi_k} \neq 0} \frac{\|z\|}{\|\xi_k\|_{\Xi_k}}, \quad k = 1, \ldots, N, \tag{4.1}$$

where the disturbance $\xi_k = \xi_k(t)$ belongs to Ξ_k, one of the given classes of signals, and $z = z_k$ is the objective output in response to the disturbances $\xi_k(t)$ and $\xi_s(t) \equiv 0$ for all $s \neq k$ under zero initial conditions, $\|z\|$ is the norm in the z-space, and $\|\cdot\|_{\Xi_k}$ is the norm in Ξ_k. The problem to be considered is to characterize Pareto optimal solutions

$$\Theta_P = \arg\min_{\Theta}\{J_k(\Theta),\ k = 1, \ldots, N\}.$$

To this goal, let us define on the trajectories of the system under disturbances acting on all inputs simultaneously the multi-objective cost

$$J_\alpha(\Theta) = \sup_{\xi_k \in \Xi_k, \forall k} \frac{\|z\|}{\sum\limits_{k=1}^{N} \alpha_k \|\xi_k\|_{\Xi_k}}, \tag{4.2}$$

where $\alpha = (\alpha_1, \ldots, \alpha_N)$, $\alpha_k > 0$, $k = 1, \ldots, N$, $\sum_{k=1}^{N} \alpha_k = 1$. Suppose that $\Theta_P \in \mathcal{P}$ and $J_k(\Theta_P) = \gamma_k > 0$, $k = 1, \ldots, N$. From (4.1) it immediately follows that $\forall\, \xi_k \in \Xi_k$, $k = 1, \ldots, N$

$$\|z\| = \|\sum_{k=1}^{N} z_k\| \leq \sum_{k=1}^{N} \gamma_k \|\xi_k\|_{\Xi_k} = \gamma \sum_{k=1}^{N} \alpha_k \|\xi_k\|_{\Xi_k},$$

where $\alpha_k = \gamma_k/\gamma$, $\gamma = \sum_{k=1}^{N} \gamma_k$. Here z_k is the objective output in response to the disturbance ξ_k from the class Ξ_k under $\xi_s = 0$ for all $s \neq k$. Consequently, $J_\alpha(\Theta_P) \leq \gamma$. Now, if Θ_α is a minimum of $J_\alpha(\Theta)$, we have $J_\alpha(\Theta_\alpha) = \gamma_\alpha \leq \gamma$. In turn, this means that

$$\|z_k\| \leq \gamma_\alpha \alpha_k \|\xi_k\|_{\Xi_k} = \gamma_\alpha \frac{\gamma_k}{\gamma} \|\xi_k\|_{\Xi_k} \leq \gamma_k \|\xi_k\|_{\Xi_k} \quad \forall\, \xi_k \in \Xi_k$$

or, in other words, $J_k(\Theta_\alpha) \leq J_k(\Theta_P)$, $k = 1, \ldots, N$. By the definition of the Pareto set, it can be satisfied only if $\Theta_\alpha \in \mathcal{P}$. Thus, we arrive at the following necessary condition for Pareto optimality.

Theorem 4.1. *Let* $(\gamma_1, \ldots, \gamma_N)$ *be a Pareto optimal point in the space of criteria and* Θ_α *minimize the multi-objective cost* $J_\alpha(\Theta)$ *under* $\alpha_i = \gamma_i / \sum_{k=1}^{N} \gamma_k$. *Then* $\Theta_\alpha \in \mathcal{P}$ *and* $J_k(\Theta_\alpha) = \gamma_k$, $k = 1, \ldots, N$.

According to this statement, Pareto optimal solutions should be sought among the optimal solutions for the performance measure $J_\alpha(\Theta)$. However, due to the fact that finding the optimal solutions for this cost is a rather complex issue, we look at another family of one-criterion problems whose solutions are easy to find and allow us to estimate the boundary of the Pareto set in the space of criteria.

Namely, we introduce an auxiliary multi-objective cost

$$\Gamma_\alpha(\Theta) = \sup_{\xi_k \in \Xi_k, \forall k} \frac{||z||}{(\sum\limits_{k=1}^{N} \alpha_k^2 ||\xi_k||_{\Xi_k}^2)^{1/2}} \tag{4.3}$$

which is the worst-case disturbance attenuation level for disturbances acting on all inputs. In the sequel, we will show the optimal solutions for this criterion can be effectively tackled for some classes of disturbances. From (4.3) it immediately follows that

$$\Gamma_\alpha(\Theta) \geq \max\{J_k(\Theta)/\alpha_k,\ k = 1, \ldots, N\}. \tag{4.4}$$

Since

$$\frac{||z||}{\sum\limits_{k=1}^{N} \alpha_k ||\xi_k||_{\Xi_k}} \geq \frac{1}{\sqrt{N}} \frac{||z||}{(\sum\limits_{k=1}^{N} \alpha_k^2 ||\xi_k||_{\Xi_k}^2)^{1/2}},$$

we have $J_\alpha(\Theta) \geq (1/\sqrt{N})\Gamma_\alpha(\Theta)$. Let $J_k(\Theta_P) = \gamma_k > 0$, $k = 1, \ldots, N$ for some $\Theta_P \in \mathcal{P}$. Since Θ_α is the minimum of $J_\alpha(\Theta)$ for $\alpha_i = \gamma_i/\gamma$ and $\gamma = \sum_{k=1}^{N} \gamma_k$, we have $J_\alpha(\Theta_\alpha) = \gamma_\alpha \leq \gamma$. Consequently, the following inequalities hold:

$$\gamma \geq J_\alpha(\Theta_\alpha) \geq (1/\sqrt{N})\Gamma_\alpha(\Theta_\alpha) \geq (1/\sqrt{N})\Gamma_\alpha(\Theta_\alpha^*)$$

where Θ_α^* is a minimum of $\Gamma_\alpha(\Theta)$. Now, in view of (4.4), we obtain for $k = 1, \ldots, N$

$$(1/\sqrt{N})J_k(\Theta_\alpha^*) \leq (1/\sqrt{N})\Gamma_\alpha(\Theta_\alpha^*)\alpha_k \leq \gamma\alpha_k = \gamma_k. \tag{4.5}$$

Since $\Theta_\alpha \in \mathcal{P}$, there exists at least one criterion $J_s(\Theta)$ such that $J_s(\Theta_\alpha^*) > J_s(\Theta_\alpha) = \gamma_s$. In view of (4.5) we obtain that $(1/\sqrt{N})J_s(\Theta_\alpha^*) \leq J_s(\Theta_P) < J_s(\Theta_\alpha^*)$ for all such indices s or, in other words,

$$\frac{J_s(\Theta_\alpha^*) - J_s(\Theta_P)}{J_s(\Theta_\alpha^*)} \leq 1 - \frac{\sqrt{N}}{N}, \tag{4.6}$$

whereas $J_k(\Theta_\alpha^*) \leq \gamma_k$ for all other indices. Thus, the relative losses on each criterion for the optimal solution with regard to the multi-objective cost $\Gamma_\alpha(\Theta)$ compared to the corresponding Pareto optimal solution, if any, do not exceed $1 - \sqrt{N}/N$.

Now, we show that the above estimates of relative losses for the optimal solutions with regard to the performance measure $\Gamma_\alpha(\Theta)$ can be improved. Indeed, suppose there exists (Θ_α^*, ξ^*), where ξ^* denotes $(\xi_1^*, \ldots, \xi_N^*)$, a solution to the problem

$$\inf_{\Theta} \sup_{\xi_k \in \Xi_k, \forall k} \frac{\|z\|}{(\sum\limits_{k=1}^{N} \alpha_k^2 \|\xi_k\|_{\Xi_k}^2)^{1/2}}. \tag{4.7}$$

We write the following inequalities:

$$\Gamma_\alpha(\Theta_\alpha^*) = \frac{\|z(\Theta_\alpha^*, \xi^*)\|}{(\sum\limits_{k=1}^{N} \alpha_k^2 \|\xi_k^*\|_{\Xi_k}^2)^{1/2}} \leq \frac{\|z(\Theta_\alpha, \xi^*)\|}{\mu(\alpha) \sum\limits_{k=1}^{N} \alpha_k \|\xi_k^*\|_{\Xi_k}},$$

where $z(\cdot, \cdot)$ is the objective output for the corresponding matrix Θ and disturbance inputs, $\Theta_\alpha \in \mathcal{P}$, $J_k(\Theta_\alpha) = \gamma_k$, $\alpha_i = \gamma_i/\gamma$, $\gamma = \sum_{k=1}^{N} \gamma_k$, and

$$\mu(\alpha) = \frac{(\sum\limits_{k=1}^{N} \alpha_k^2 \|\xi_k^*\|_{\Xi_k}^2)^{1/2}}{\sum\limits_{k=1}^{N} \alpha_k \|\xi_k^*\|_{\Xi_k}}, \quad \frac{1}{\sqrt{N}} \leq \mu(\alpha) \leq 1. \tag{4.8}$$

This implies

$$\mu(\alpha)\Gamma_\alpha(\Theta_\alpha^*) \leq \frac{\|z(\Theta_\alpha, \xi^*)\|}{\sum\limits_{k=1}^{N} \alpha_k \|\xi_k^*\|_{\Xi_k}} \leq \sup_{\xi_k \in \Xi_k, \forall k} \frac{\|z(\Theta_\alpha, \xi)\|}{\sum\limits_{k=1}^{N} \alpha_k \|\xi_k\|_{\Xi_k}} = J_\alpha(\Theta_\alpha) \leq \gamma,$$

where $\xi = (\xi_1, \ldots, \xi_N)$. Therefore, in view of (4.4),

$$\mu(\alpha) J_k(\Theta_\alpha^*) \le \alpha_k \gamma = \gamma_k, \quad k = 1, \ldots, N.$$

Thus, we can formulate the following main result.

Theorem 4.2. *The relative losses on each criterion for the optimal solution Θ_α^* with regard to the multi-objective cost $\Gamma_\alpha(\Theta)$ compared to the corresponding Pareto optimal solution, if any, for which $J_k(\Theta_P) = \gamma_k$, $k = 1, \ldots, N$ and $\alpha_i = \gamma_i / \sum_{k=1}^N \gamma_k$, do not exceed $1 - \mu(\alpha)$, i.e.*

$$\frac{J_k(\Theta_\alpha^*) - J_k(\Theta_P)}{J_k(\Theta_\alpha^*)} \le 1 - \mu(\alpha), \quad k = 1, \ldots, N, \tag{4.9}$$

where $1/\sqrt{N} \le \mu(\alpha) \le 1$ is defined in (4.8).

5. Extension for Possible Coalitions of Disturbances

The above characterization of the Pareto optimal set also allows one to synthesize the Pareto suboptimal solutions in problems with multiple disturbance coalitions formed from disturbances acting into different inputs. Specifically, suppose the disturbance inputs $\xi_1, \ldots, \xi_N$ in the system are combined in S coalitions so that each new disturbance input η_l, $l = 1, \ldots, S$ consists of some disturbance inputs from the set $\{\xi_1, \ldots, \xi_N\}$. Note that the same disturbance input ξ_k may enter at various coalitions. In this case, such a disturbance input must be included in the set $\{\xi_1, \ldots, \xi_N\}$ corresponding number of times under different numbers (see the final example). Assume that ξ_k, $k = 1, \ldots, N$ enters into the coalition η_l with $l = f(k)$, where function $f(\cdot)$ maps the set $\{1, \ldots, N\}$ onto the set $\{1, \ldots, S\}$. Associated to this decomposition are defined the following disturbance attenuation criteria to be minimized:

$$\hat{J}_l(\Theta) = \sup_{\|\eta_l\|_{\Pi_l} \neq 0} \frac{\|z\|}{\|\eta_l\|_{\Pi_l}}, \quad l = 1, \ldots, S, \tag{5.1}$$

where Π_l is the direct production of the Ξ_k-spaces for the signals within the given coalition,

$$\|\eta_l\|_{\Pi_l}^2 = \sum_{\{k: f(k) = l\}} \rho_k^2 \|\xi_k\|_{\Xi_k}^2, \quad l = 1, \ldots, S,$$

and $\rho_k > 0$, $k = 1, \ldots, N$ are weighting coefficients. The problem is to characterize Pareto optimal solutions in this multi-objective disturbance attenuation problem under disturbance coalitions, i.e.

$$\Theta_P = \arg\min_{\Theta}\{\hat{J}_l(\Theta),\ l = 1, \ldots, S\}. \tag{5.2}$$

This problem can be also regarded as a game between the nature which chooses one of the S possible coalitions of the disturbances and a person choosing the matrix Θ in such a way to minimize his losses for each of such coalitions.

By applying the similar arguments as those in the case without coalitions, we find the optimal and suboptimal multi-objective costs

$$\begin{aligned}
\hat{J}_\alpha(\Theta) &= \sup_{\eta_l \in \Pi_l, \forall\, l} \frac{\|z\|}{\sum\limits_{l=1}^{S} \alpha_l \|\eta_l\|_{\Pi_l}} = \sup_{\xi_k \in \Pi_k, \forall\, k} \frac{\|z\|}{\sum\limits_{l=1}^{S} \alpha_l \left(\sum\limits_{\{k: f(k)=l\}} \rho_k^2 \|\xi_k\|^2_{\Xi_k} \right)^{1/2}}, \\
\hat{\Gamma}_\alpha(\Theta) &= \sup_{\eta_l \in \Pi_l, \forall\, l} \frac{\|z\|}{(\sum\limits_{l=1}^{S} \alpha_l^2 \|\eta_l\|^2_{\Pi_l})^{1/2}} = \sup_{\xi_k \in \Xi_k, \forall\, k} \frac{\|z\|}{(\sum\limits_{k=1}^{N} \alpha^2_{f(k)} \rho^2_{f(k)} \|\xi_k\|^2_{\Xi_k})^{1/2}},
\end{aligned} \tag{5.3}$$

where $\alpha = (\alpha_1, \ldots, \alpha_S)$, $\alpha_l > 0$, $l = 1, \ldots, S$, $\sum_{l=1}^{S} \alpha_l = 1$, for which the statements formulated in Theorems 4.1 and 4.2 hold true. The relative losses on each criterion for the optimal solution Θ^*_α with regard to the suboptimal multi-objective cost $\hat{\Gamma}_\alpha(\Theta)$ compared to the corresponding Pareto optimal solution, if any, for which $\hat{J}_l(\Theta_P) = \gamma_l$, $l = 1, \ldots, S$ and $\alpha_l = \gamma_l / \sum_{l=1}^{S} \gamma_l$, do not exceed $1 - \hat{\mu}(\alpha)$, i.e.

$$\frac{\hat{J}_l(\Theta^*_\alpha) - \hat{J}_l(\Theta_P)}{\hat{J}_l(\Theta^*_\alpha)} \le 1 - \hat{\mu}(\alpha), \quad l = 1, \ldots, S, \tag{5.4}$$

where

$$\hat{\mu}(\alpha) = \frac{(\sum\limits_{l=1}^{S} \alpha_l^2 \|\eta_l^*\|^2_{\Pi_l})^{1/2}}{\sum\limits_{l=1}^{S} \alpha_l \|\eta^*_l\|_{\Pi_l}} = \frac{(\sum\limits_{k=1}^{N} \alpha^2_{f(k)} \rho^2_{f(k)} \|\xi^*_k\|^2_{\Xi_k})^{1/2}}{\sum\limits_{l=1}^{S} \alpha_l \left(\sum\limits_{\{k: f(k)=l\}} \rho_k^2 \|\xi^*_k\|^2_{\Xi_k} \right)^{1/2}}, \quad \frac{1}{\sqrt{S}} \le \hat{\mu}(\alpha) \le 1, \tag{5.5}$$

$(\eta^*_1, \ldots, \eta^*_S)$ is a worst-case disturbance with respect to the performance measure $\hat{\Gamma}_\alpha(\Theta)$ given in (5.3) and $\eta^*_l = \{\xi^*_k, \forall\, k : f(k) = l\}$.

Thus, we have established a special role of the performance measures $\Gamma_\alpha(\Theta)$ and $\hat{\Gamma}_\alpha(\Theta)$ in the multi-objective attenuation problems without and with coalitions of disturbances. The optimal solutions with respect to each of these multi-objective costs are the Pareto suboptimal solutions in the corresponding problem. Now, the above approach can be applied to any control or filtering multi-objective problem in which the Pareto suboptimal solutions and their relative losses are effectively computed.

6. Pareto Suboptimal Solutions for H_∞ and/or γ_0 Criteria in Continuous-Time Systems

6.1. H_∞/γ_0-Norm

Consider an internally stable linear time-invariant system parameterized by the parameter matrix Θ and described by the equations

$$\begin{aligned} \dot{x} &= A(\Theta)x + \sum_{i=1}^{m} B_i(\Theta)w_i + \sum_{j=m+1}^{N} B_j(\Theta)w_j, \quad x(0) = 0, \\ z &= C(\Theta)x + \sum_{i=1}^{m} D_i(\Theta)w_i, \end{aligned} \tag{6.1}$$

where $x \in R^{n_x}$ is the state, $z \in R^{n_z}$ is the objective output. Suppose that the disturbances acting on inputs w_i, $i = 1, \ldots, m$ are L_2-norm bounded signals while the disturbances acting on inputs w_j, $j = m+1, \ldots, N$ are impulsive signals of the form $w_j = d_j\delta(t)$, where $\delta(t)$ is the Dirac δ-function and d_j are unknown vectors. From (2.2) and (2.7) it follows that the disturbance attenuation levels for such signals coincide with the corresponding criteria under consideration.

According to Theorem 4.2, the suboptimal multi-objective cost defined in (4.3) takes the form

$$\Gamma_\alpha(\Theta) = \sup_{w_i \in L_2\,\forall\, i,\, d_j \in R^{n_j}\,\forall\, j} \frac{\|z\|}{(\sum_{i=1}^{m} \alpha_i^2\|w_i\|^2 + \sum_{j=m+1}^{N} \alpha_j^2|d_j|^2)^{1/2}}. \tag{6.2}$$

We will show that Γ_α-optimal solutions for the system (6.1), called by the Pareto suboptimal solutions, minimize a disturbance attenuation level in an auxiliary system with two disturbance inputs, one of which for an L_2-disturbance and another for an impulsive disturbance.

Indeed, let us write the equations of the system (6.1) in the following form:

$$\begin{aligned} \dot{x} &= \mathcal{A}x + \mathcal{B}_v v + \mathcal{B}_\zeta \zeta, \quad x(0) = 0, \\ z &= \mathcal{C}x + \mathcal{D}_v v, \end{aligned} \tag{6.3}$$

where $v = \text{col}(v_1, \ldots, v_m) \in L_2$, $\zeta = \hat{d}\delta(t)$, $\hat{d} = \text{col}(\hat{d}_{m+1}, \ldots, \hat{d}_N) \in R^{\hat{n}}$, $\hat{n} = \sum_{j=m+1}^{N} n_j$,

$$\begin{aligned} \mathcal{A} = A(\Theta), \quad \mathcal{B}_v = (\alpha_1^{-1}B_1(\Theta)\ldots\alpha_m^{-1}B_m(\Theta)), \quad \mathcal{B}_\zeta = (\alpha_{m+1}^{-1}B_{m+1}(\Theta)\ldots\alpha_N^{-1}B_N(\Theta)), \\ \mathcal{C} = C(\Theta), \quad \mathcal{D}_v = (\alpha_1^{-1}D_1(\Theta)\ldots\alpha_m^{-1}D_m(\Theta)). \end{aligned} \tag{6.4}$$

It is clear that for $v_i = \alpha_i w_i$, $i = 1, \ldots, m$ and $\hat{d}_j = \alpha_j d_j$, $j = m+1, \ldots, N$, the equations (6.1) and (6.3) are the same, whereas the performance measure (6.2) takes the form (for simplicity, argument Θ is omitted)

$$\Gamma_\alpha = \sup_{v \in L_2, \hat{d} \in R^{\hat{n}}} \frac{\|z\|}{(\|v\|^2 + |\hat{d}|^2)^{1/2}}. \tag{6.5}$$

On one hand, this is the disturbance attenuation level for L_2-disturbances $v(t)$ and impulsive disturbances $\zeta = \hat{d}\delta(t)$ in the system (6.3) as the worst-case norm of the objective output over all admissible disturbances acting in two channels with transfer matrices H_v and H_ζ. On the other hand, this is the induced norm of the linear operator generated by the system (6.3), which maps the pair $(v(t), \hat{d})$ to $z(t)$. This norm can be called the H_∞/γ_0-norm and denoted by $\gamma_{\infty,0}$. It is clear that

$$\gamma_{\infty,0} \geq \max\{\gamma_\infty, \gamma_0\},$$

where $\gamma_\infty = \|H_v\|_\infty$ and $\gamma_0 = \gamma_0(H_\zeta)$.

Since system (6.3) under impulsive disturbances can be equivalently presented by the equations

$$\begin{aligned} \dot{x} &= \mathcal{A}x + \mathcal{B}_v v, \\ z &= \mathcal{C}x + \mathcal{D}_v v \end{aligned} \tag{6.6}$$

with an unknown initial state of the form $x(0) = \mathcal{B}_\zeta \hat{d}$, we arrive at the so called problem of the H_∞ norm with transients, for which the performance measure

$$\gamma_v = \sup_{\|v\|^2 + x_0^T R x_0 \neq 0} \frac{\|z\|}{(\|v\|^2 + x_0^T R x_0)^{1/2}} \tag{6.7}$$

was introduced in [18]. This measure has been characterized in [21] in terms of LMIs as follows: $\gamma_v < \gamma$ if and only if there exists a $X = X^T > 0$ such that

$$\begin{pmatrix} \mathcal{A}^T X + X\mathcal{A} & \star & \star \\ \mathcal{B}_v^T X & -\gamma^2 I & \star \\ \mathcal{C} & \mathcal{D}_v & -I \end{pmatrix} < 0, \quad X < \gamma^2 R \tag{6.8}$$

are feasible. Note that the first inequality in (6.8) implies that the derivative of the function $V(x) = x^T X x$ along a trajectory of the system (6.6) satisfies inequality

$$\dot{V} + |z|^2 - \gamma^2 |v|^2 < 0 \quad \forall x, v\, (|x|^2 + |v|^2 \neq 0). \tag{6.9}$$

Integrating this inequality over infinite horizon, we get $\|z\|^2 - \gamma^2\|v\|^2 < x_0^T X x_0$. Then the second inequality in (6.8) implies $x_0^T X x_0 < \gamma^2 x_0^T R x_0$ for all x_0 and hence $\gamma_v < \gamma$. In addition, it has been also shown that, if the initial state is known to belong to a linear manifold of the form $x = T\hat{d}$ with matrix T being of a full column rank, the inequality

$$\frac{\|z\|}{(\|v\|^2 + x_0^T R x_0)^{1/2}} < \gamma \quad \forall v \in L_2,\ \forall x_0 = T\hat{d}, \quad \hat{d} \in R^{n_d}$$

holds if and only if the first inequality in (6.8) and $T^T X T < \gamma^2 T^T R T$ are feasible.

Returning to the characterization of the performance measure (6.5) for the system (6.3) under $x(0) = \mathcal{B}_\xi \hat{d}$ and choosing $T = \mathcal{B}_\zeta$, $R = \rho^2 \mathcal{B}_\zeta (\mathcal{B}_\zeta^T \mathcal{B}_\zeta)^{-1} \mathcal{B}_\zeta^T$, we immediately come to the following result.

Theorem 6.1. *The H_∞/γ_0 norm $\gamma_{\infty,0}$ for the system (6.3) is less than a prescribed level γ in the system (6.3) if and only if there exists a $(n_x \times n_x)$-matrix $X = X^T > 0$ such that LMIs*

$$\begin{pmatrix} \mathcal{A}^T X + X\mathcal{A} & \star & \star \\ \mathcal{B}_v^T X & -\gamma^2 I & \star \\ \mathcal{C} & \mathcal{D}_v & -I \end{pmatrix} < 0, \quad \mathcal{B}_\zeta^T X \mathcal{B}_\zeta < \gamma^2 I \tag{6.10}$$

are feasible.

Note that $\gamma_{\infty,0}$ can be computed as the minimal value of γ^2 for which LMIs (6.10) are feasible in variables $X = X^T > 0$ and $\gamma^2 > 0$. Besides, H_∞-norm of the transfer matrix $H_v(s) = \mathcal{C}(sI - \mathcal{A})^{-1}\mathcal{B}_v + \mathcal{D}_v$ can be computed as a minimum value of γ^2 for which the first inequality in (6.10) is feasible, whereas γ_0-norm of the transfer matrix $H_\zeta(s) = \mathcal{C}(sI - \mathcal{A})^{-1}\mathcal{B}_\zeta$ can be obtained, in view of (2.8), as a minimum value of γ^2, for which the first inequality in (6.10), where the second block row and column are deleted, and the second inequality in (6.10) should be feasible.

Remark 6.1. *It is important to highlight that if X_∞ is the solution to the first inequality in (6.10) at the minimal value of $\gamma_\infty^2 = \|H_v\|_\infty^2$ and $\mathcal{B}_\zeta^T X_\infty \mathcal{B}_\zeta < \gamma_\infty^2 I$, then $\gamma_{\infty,0} = \gamma_\infty$. This means that the H_∞/γ_0 norm is the trade-off between the H_∞ norm of the transfer matrix H_v and the γ_0 norm of the transfer matrix H_ζ if and only if the following inequality holds*

$$\lambda_{max}(\mathcal{B}_\zeta^T X_\infty \mathcal{B}_\zeta) \geq \gamma_\infty^2. \tag{6.11}$$

Now, let us find the worst-case disturbances $v^*(t) \in L_2$ and $\zeta^*(t) = \hat{d}^*\delta(t)$ with respect to the performance measure (6.5). We will show that, in the case when $\gamma_{\infty,0} > \gamma_\infty$, the supremum in (6.5) is actually achieved.

Theorem 6.2. *Suppose the inequality (6.11) holds. Then the H_∞/γ_0 norm takes its maximal value γ_α under the worst-case disturbance*

$$v^*(t) = L_* x^*(t),\ L_* = (\gamma_\alpha^2 I - \mathcal{D}^T\mathcal{D})^{-1}(\mathcal{C}^T\mathcal{D} + X_\gamma \mathcal{B}_v)^T \tag{6.12}$$

and $\zeta^*(t) = \hat{d}^*\delta(t)$ with $\hat{d}^* = e_{max}(\mathcal{B}_\zeta^T X_\gamma \mathcal{B}_\zeta)$, where $X_\gamma = X_\gamma^T > 0$ is the stabilizing solution of the Riccati equation

$$Ric(X,\gamma) = \mathcal{A}^T X + X\mathcal{A} + \mathcal{C}^T\mathcal{C} + (\mathcal{C}^T\mathcal{D} + X\mathcal{B}_v)(\gamma^2 I - \mathcal{D}^T\mathcal{D})^{-1}(\mathcal{C}^T\mathcal{D} + X\mathcal{B}_v)^T = 0 \tag{6.13}$$

with $\gamma = \gamma_\alpha$, $x_*(t)$ is the solution of the system

$$\dot{x} = (\mathcal{A} + \mathcal{B}_v L_*)x + \mathcal{B}_\zeta \zeta^*, \quad x(0) = 0$$

and $e_{max}(\mathcal{B}_\zeta^T X_\gamma \mathcal{B}_\zeta)$ is the normalized eigenvector associated with the maximum eigenvalue γ_α^2 of the matrix $\mathcal{B}_\zeta^T X_\gamma \mathcal{B}_\zeta$.

Proof. According to Remark 6.1 $\gamma_\infty < \gamma_\alpha$, hence, for the linear feedback disturbance v^* defined in (6.12), the equality $\|z\|^2 - \gamma_\alpha^2\|v^*\|^2 = \hat{d}^T\mathcal{B}_\zeta^T X_\gamma \mathcal{B}_\zeta \hat{d}$ holds (see, for example, [26]). Then we have

$$\frac{\|z\|^2}{\|v^*\|^2 + |\hat{d}|^2} = \gamma_\alpha^2 + \frac{\hat{d}^T(\mathcal{B}_\zeta^T X_\gamma \mathcal{B}_\zeta - \gamma_\alpha^2 I)\hat{d}}{\|v^*\|^2 + |\hat{d}|^2}.$$

In view of

$$\frac{\|z\|^2}{\|v^*\|^2 + |\hat{d}|^2} \le \gamma_\alpha^2, \tag{6.14}$$

we get $\mathcal{B}_\zeta^T X_\gamma \mathcal{B}_\zeta \le \gamma_\alpha^2 I$.

Now, we will show that there exists $\hat{d}^*$ such that $\hat{d}^{*T}(\mathcal{B}_\zeta^T X_\gamma \mathcal{B}_\zeta - \gamma_\alpha^2 I)\hat{d}^* = 0$. Suppose that $\mathcal{B}_\zeta^T X_\gamma \mathcal{B}_\zeta < \gamma_\alpha^2 I$. Then, for sufficiently small $\varepsilon > 0$, there exists $X(\varepsilon)$ that is a stabilizing solution of the equation $Ric(X, \gamma_\alpha) + \varepsilon I = 0$ such that $\mathcal{B}_\zeta^T X_\gamma \mathcal{B}_\zeta < \mathcal{B}_\zeta^T X(\varepsilon)\mathcal{B}_\zeta < \gamma_\alpha^2 I$. Consequently, $X(\varepsilon)$ satisfies both LMIs in (6.10), which contradicts to the definition of Γ_α. This means that for $\hat{d}^* = e_{max}(\mathcal{B}_\zeta^T X_\gamma \mathcal{B}_\zeta)$, the inequality (6.14) becomes the equality.

Corollary 6.1. *Let the inequality (6.11) hold. Then* $\gamma_{\infty,0}^2 = \lambda_{max}(\mathcal{B}_\zeta^T X_\gamma \mathcal{B}_\zeta)$, *where* $X_\gamma = X_\gamma^T > 0$ *is the stabilizing solution of the Riccati equation (6.13) for* $\gamma = \gamma_{\infty,0}$, *and it can be computed as* $\gamma_{\infty,0}^2 = \min \gamma^2$ *subject to the non-strict LMIs (6.10).*

6.2. State-Feedback Controllers

Consider a system described by the equations

$$\begin{aligned} \dot{x} &= Ax + \sum_{i=1}^{m} B_i w_i + \sum_{j=m+1}^{N} B_j w_j + B_u u, \\ z &= C_z x + \sum_{i=1}^{m} D_{z,i} w_i + D_u u, \quad x(0) = 0, \end{aligned} \tag{6.15}$$

where $x \in R^{n_x}$ is the state, $u \in R^{n_u}$ is the control input, $z \in R^{n_z}$ is the controlled output, $w_i \in R^{n_i}$, $i = 1, \ldots, m$ are L_2-norm bounded signals and $w_j \in R^{n_j}$, $j = m+1, \ldots, N$ are impulsive signals of the form $w_j = d_j\delta(t)$, where d_j, $j = m+1, \ldots, N$ are unknown vectors. The goal is to synthesize the Pareto suboptimal state-feedback control laws $u = \Theta x$ in the multi-objective problem

$$\Theta_P = \arg\min_{\Theta}\{\|H_i(\Theta)\|_\infty, \forall i;\ \gamma_0(H_j(\Theta)), \forall j\}. \tag{6.16}$$

The characterization of $\Gamma_\alpha(\Theta)$ obtained above can be immediately applied to the system of the form (6.3) with matrices given in (6.4), where $A(\Theta) = A + B_u\Theta$, $C(\Theta) = C_z + D_u\Theta$, $B_k(\Theta) = B_k$, $k = 1,\ldots,N$, $D_i(\Theta) = D_{z,i}$, $i = 1,\ldots,m$. By substituting the matrices of the closed-loop system into LMIs (6.10), multiplying the first inequality by diag (X^{-1}, I, I) from the left and from the right, introducing the new variables $Y = X^{-1}$ and $Z = \Theta Y$, and changing the second inequality by using Schur lemma, we immediately arrive at the following.

Theorem 6.3. *The gain matrix of Pareto suboptimal state-feedback controller for the system (6.15) is computed as* $\Theta_\alpha^* = Z_\alpha Y_\alpha^{-1}$, *where* $(Y_\alpha, Z_\alpha, \gamma_\alpha^2)$ *is a solution of the problem* $\inf \gamma^2 = \gamma_\alpha^2$ *subject to constraints*

$$\begin{pmatrix} AY + YA^T + B_uZ + Z^TB_u^T & \star & \cdots & \cdots & \star \\ B_1^T & -\alpha_1^2\gamma^2 I & \star & \cdots & \star \\ \cdots & \cdots & \cdots & \cdots & \cdots \\ B_m^T & 0 & \cdots & -\alpha_m^2\gamma^2 I & \star \\ C_zY + D_uZ & D_{z,1} & \cdots & D_{z,m} & -I \end{pmatrix} < 0,$$

$$\begin{pmatrix} Y & \star & \cdots & \star \\ B_{m+1}^T & \alpha_{m+1}^2\gamma^2 I & \star & \star \\ \cdots & \cdots & \cdots & \cdots \\ B_N^T & 0 & \cdots & \alpha_N^2\gamma^2 I \end{pmatrix} > 0. \tag{6.17}$$

Remark 6.2. *In the particular case when* $C_z^TD_u = 0$, $D_u^TD_u = I$ *and* $D_{z,i} = 0$, $i = 1,\ldots,m$ *in view of Corollary 6.1 we obtain that* $\Theta_\alpha^* = -B_u^TX_*$ *and* $\gamma_\alpha^2 = \lambda_{max}(\mathcal{B}_\varsigma^TX_*\mathcal{B}_\varsigma)$, *where* $X_* = X_*^T > 0$ *is the stabilizing solution of the Riccati equation*

$$A^TX + XA + C_z^TC_z + X(\gamma_\alpha^{-2}\mathcal{B}_v\mathcal{B}_v^T - B_uB_u^T)X = 0, \tag{6.18}$$

where

$$\mathcal{B}_v = (\alpha_1^{-1}B_1\ldots\alpha_m^{-1}B_m), \quad \mathcal{B}_\varsigma = (\alpha_{m+1}^{-1}B_{m+1}\cdots\alpha_N^{-1}B_N).$$

According to Theorem 4.2, the relative losses on each criterion for the Γ_α-optimal controllers do not exceed $1 - \mu(\alpha)$, where

$$\mu(\alpha) = \frac{(\sum\limits_{i=1}^{m} \alpha_i^2 \|w_i^*\|^2 + \sum\limits_{j=m+1}^{N} \alpha_j^2 |d_j^*|^2)^{1/2}}{\sum\limits_{i=1}^{m} \alpha_i \|w_i^*\| + \sum\limits_{j=m+1}^{N} \alpha_j |d_j^*|} \tag{6.19}$$

and $w_1^*(t), \dots, w_m^*(t), d_{m+1}^*, \dots, d_N^*$ form the worst-case disturbance for the Γ_α-optimal closed-loop system. These relative losses can be equivalently expressed in terms of the worst-case disturbance for the system (6.3), (6.4) with $\Theta = \Theta_\alpha^*$ as follows

$$\mu(\alpha) = \frac{\sqrt{\|v^*\|^2 + 1}}{\sum\limits_{i=1}^{m} \|v_i^*\| + \sum\limits_{j=m+1}^{N} |\hat{d}_j^*|}, \tag{6.20}$$

where $v^* = \mathrm{col}(v_1^*, \dots, v_m^*)$, $v^*(t) = L_* x_*(t)$, $v_i^* = L_i x_*(t)$, $L_* = (L_1, \dots, L_m)$, $\hat{d}^* = \mathrm{col}\,(d_{m+1}^*, \dots, d_N^*)$, $\hat{d}^* = e_{max}(\mathcal{B}_\zeta^T X_\gamma \mathcal{B}_\zeta)$, $(|\hat{d}^*| = 1)$ are determined in Theorem 6.2 provided that $\mathcal{A}$ is replaced by $A + B_u \Theta_\alpha^*$. Note that $\|v_i^*\|^2 = \hat{d}^{*T} \mathcal{B}_\zeta^T X_i \mathcal{B}_\zeta \hat{d}^*$, where X_i is a solution to the equation

$$(A + \mathcal{B}_v L_* + B_u \Theta_\alpha^*)^T X + X(A + \mathcal{B}_v L_* + B_u \Theta_\alpha^*) + L_i^T L_i = 0.$$

It is interesting to compare the set of the Pareto suboptimal state-feedback controllers obtained for two criteria, H_∞ and γ_0 norms, with the set of controllers synthesized on the basis of the mixed H_2/H_∞ performance measure. For simplicity of the presentation we assume that $D_{z,1} = 0$, $C_z^T D_u = 0$ and $D_u^T D_u = I$. In view of Theorem 6.3 and Remark 6.2, the gain matrix of the Γ_α-controller is computed as $\Theta_\alpha^* = -B_u^T X_\alpha$, where $X_\alpha = X_\alpha^T > 0$ is a stabilizing solutions to the equation

$$A^T X + XA + C_z^T C_z + X[(\alpha\gamma_\alpha)^{-2} B_1 B_1^T - B_u^T B_u] X = 0, \tag{6.21}$$

and $\gamma_\alpha^2 = (1-\alpha)^{-2} \lambda_{max}(B_2^T X_\alpha B_2)$.

Recall that the mixed H_2/H_∞ performance measure of the closed-loop system for a given $\nu > 0$ is determined as tr$\,(B_2^T X B_2)$, where $X = X^T > 0$ is a stabilizing solution to the Riccati equation

$$\mathcal{A}^T X + X\mathcal{A} + \mathcal{C}^T \mathcal{C} + \nu^{-2} X B_1 B_1^T X = 0, \tag{6.22}$$

where $\mathcal{A} = A + B_u\Theta$, $\mathcal{C} = C_z + D_u\Theta$. The gain matrix of the H_2/H_∞ controller minimizing the mixed H_2/H_∞ performance measure for a given ν is computed as $\Theta_\nu = -B_u^T X_\nu$, where $X_\nu = X_\nu^T > 0$ is a stabilizing solutions to the equation

$$A^T X + XA + C_z^T C_z + X(\nu^{-2} B_1 B_1^T - B_u^T B_u)X = 0. \quad (6.23)$$

It is obvious that when $\alpha\gamma_\alpha = \nu$, i.e. when $\alpha = \nu/[\nu + \lambda_{max}^{1/2}(B_2^T X_\nu B_2)] < 1$, we get $\Theta_\alpha = \Theta_\nu$.

In the case when all criteria in (6.16) are H_∞ norms, the disturbance attenuation level for disturbances acting in all channels takes the form

$$\Gamma_\alpha(\Theta) = \sup_{w_k \in L_2, \forall k} \frac{\|z\|}{(\sum_{k=1}^{N} \alpha_k^2 \|w_k\|^2)^{1/2}} = \sup_{v \in L_2, \|v\| \neq 0} \frac{\|z\|}{\|v\|} \quad (6.24)$$

and, therefore, it is the H_∞ norm of the transfer matrix $H_{zv}(s)$ from v to z of the system (6.3), which is characterized by the first inequality in (6.10). The Pareto suboptimal state-feedback controllers are given by the formulae $u = \Theta_\alpha^* x$, where $\Theta_\alpha^* = Z_* Y_*^{-1}$ and the pair of matrices (Y_*, Z_*) is a solution to the problem $\inf \gamma^2$ subject to the first inequality in (6.17) with $m = N$. The relative losses on each criterion for the Pareto suboptimal controllers do not exceed $1 - \mu(\alpha)$, where

$$\mu(\alpha) = \lim_{\varepsilon \to 0} \frac{(\sum_{i=1}^{N} \alpha_i^2 \|w_{i\varepsilon}^*\|^2)^{1/2}}{\sum_{i=1}^{N} \alpha_i \|w_{i\varepsilon}^*\|} = \lim_{\varepsilon \to 0} \frac{\|v_\varepsilon\|}{\sum_{i=1}^{N} \|v_{i\varepsilon}\|} \quad (6.25)$$

and $v_\varepsilon(t)$ approaches the worst-case disturbance in (6.24). To compute $\|v_{i\varepsilon}\|$ we artificially introduce the impulsive signal $\zeta(t) = d\delta(t)$ with an arbitrary d and choose $\mathcal{B}_\zeta = \nu I$ with ν being such that the corresponding inequality (6.11) would be invalid, i.e. $\nu^2 \lambda_{max}(X_\infty) < \|H_v\|_\infty^2$. Then we compute the disturbance $v_\varepsilon(t)$ of the form (6.12) with $\gamma_* = \gamma_\infty + \varepsilon$ under sufficiently small $\varepsilon > 0$.

If all the criteria in (6.16) are γ_0 norms of transfer matrices $H_k(s)$, $k = 1, \ldots, N$, the disturbance attenuation level for disturbances acting in all channels takes the form

$$\Gamma_\alpha(\Theta) = \sup_{d_k \in R^{n_k}, \forall k} \frac{\|z\|}{(\sum_{k=1}^{N} \alpha_k^2 |d_k|^2)^{1/2}} = \sup_{\hat{d} \in R^{n_d}, \hat{d} \neq 0} \frac{\|z\|}{|\hat{d}|} \tag{6.26}$$

and it is the γ_0 norm of the transfer matrix $H_{z\zeta}(s)$ from ζ to z of the system (6.3), which is characterized by the LMIs in (6.10), where the second block row and column are excluded in the first inequality and $\rho = 1$ in the second inequality. The optimal state-feedback gain Θ_0 for the γ_0 norm coincides with the optimal gain for the H_2 norm and does not depend on the disturbance input matrix B_j. It is easy to see when $C_z^T D_u = 0$ and D_u is of the full column rank. By completing the square in

$$(A + B_u\Theta)^T X + X(A + B_u\Theta) + (C_z + D_u\Theta)^T(C_z + D_u\Theta) = 0 \tag{6.27}$$

and taking into account the property of monotonicity for solutions of the Lyupunov equation, we obtain that $\lambda_{max}(B_j^T X_0 B_j)$ and $\mathrm{tr}\,(B_j^T X_0 B_j)$ take their minimal values under $\Theta_0 = -(D_u^T D_u)^{-1} B_u^T X_*$, where $X_* = X_*^T > 0$ is the stabilizing solution to the Riccati equation

$$A^T X + XA + C^T C - XB_u(D_u^T D_u)^{-1} B_u^T X = 0.$$

Thus, there is a single solution to the multi-criteria problem in the case under consideration.

Theorem 6.4. *Pareto optimal state-feedback controller for the system (6.15) in the multi-objective problem with multiple γ_0 criteria is determined by the formulae $u = \Theta_\alpha^* x$, where $\Theta_\alpha^* = Z_* Y_*^{-1}$ and the pair of matrices (Y_*, Z_*) is a solution to the problem* $\inf \gamma^2$ *subject to LMIs*

$$\begin{pmatrix} AY + YA^T + B_u Z + Z^T B_u^T & \star \\ C_z Y + D_u Z & -I \end{pmatrix} < 0, \quad \begin{pmatrix} Y & \star & \dots & \star \\ B_1^T & \alpha_1^2\gamma^2 I & \star & \star \\ \dots & \dots & \dots & \dots \\ B_N^T & 0 & \dots & \alpha_N^2\gamma^2 I \end{pmatrix} > 0. \tag{6.28}$$

6.3. Output-Feedback Controllers

Consider a system described by the equations

$$
\begin{aligned}
\dot{x} &= Ax + \sum_{i=1}^{m} B_i w_i + \sum_{j=m+1}^{N} B_j w_j + B_u u, \\
z &= C_z x + \sum_{i=1}^{m} D_{z,i} w_i + D_u u, \\
y &= C_y x + \sum_{i=1}^{m} D_{y,i} w_i, \quad x(0) = 0,
\end{aligned}
\tag{6.29}
$$

where $x \in R^{n_x}$ is the state, $u \in R^{n_u}$ is the control input, $z \in R^{n_z}$ is the controlled output, $y \in R^{n_y}$ is the measurable output, $w_i \in R^{n_i}$, $i = 1, \ldots, m$ are L_2-norm bounded signals and $w_j \in R^{n_j}$, $j = m+1, \ldots, N$ are impulsive signals of the form $w_j = d_j \delta(t)$, where d_j, $j = m+1, \ldots, N$ are unknown vectors.

In the output-feedback case, the procedure of computing the parameters

$$
\Theta = \begin{pmatrix} A_r & B_r \\ C_r & D_r \end{pmatrix}
$$

of the Pareto suboptimal controllers in the form of the dynamic output-feedback

$$
\begin{aligned}
\dot{x}_r &= A_r x_r + B_r y, \quad x_r(0) = 0, \\
u &= C_r x_r + D_r y
\end{aligned}
\tag{6.30}
$$

is based on Theorem 6.1 and standard manipulations using LMI technique (see, for example, [27, 28, 21]). To this end the equations (6.29) are rewritten in the form

$$
\begin{aligned}
\dot{x} &= Ax + \mathcal{B}_v v + \mathcal{B}_\zeta \zeta + B_u u, \quad x(0) = 0, \\
z &= C_z x + \mathcal{D} v + D_u u, \\
y &= C_y x + \mathcal{D}_y v,
\end{aligned}
\tag{6.31}
$$

where

$$
\mathcal{B}_v = (\alpha_1^{-1} B_1 \ldots \alpha_m^{-1} B_m), \quad \mathcal{B}_\zeta = (\alpha_{m+1}^{-1} B_{m+1} \ldots \alpha_N^{-1} B_N),
$$

$$
\mathcal{D} = (\alpha_1^{-1} D_{z,1} \ldots \alpha_m^{-1} D_{z,m}), \quad \mathcal{D}_y = (\alpha_1^{-1} D_{y,1} \ldots \alpha_m^{-1} D_{y,m}).
$$

The closed-loop system (6.31), (6.30) is described by the equations

$$
\begin{aligned}
\dot{x}_c &= \mathcal{A}_c x_c + \mathcal{B}_{cv} v + \mathcal{B}_{c\zeta} \zeta, , \quad x_c(0) = 0, \\
z &= \mathcal{C}_c x_c + \mathcal{D}_c v,
\end{aligned}
\tag{6.32}
$$

where $x_c = \mathrm{col}\,(x, x_r)$, $\zeta = \hat{d}\delta(t)$, $\hat{d} = \mathrm{col}(d_{m+1}, \ldots, d_N)$

$$\mathcal{A}_c = \begin{pmatrix} A + B_u D_r C_y & B_u C_r \\ B_r C_y & A_r \end{pmatrix}, \quad \mathcal{B}_{c\zeta} = \begin{pmatrix} \mathcal{B}_\zeta \\ 0 \end{pmatrix}, \quad \mathcal{B}_{cv} = \begin{pmatrix} \mathcal{B}_v + B_u D_r \mathcal{D}_y \\ B_r \mathcal{D}_y \end{pmatrix},$$
$$\mathcal{C}_c = (C_z + D_u D_r C_y \quad D_u C_r), \quad \mathcal{D}_c = \mathcal{D} + D_u D_r \mathcal{D}_y. \tag{6.33}$$

By inserting these matrices into (6.10) and applying the standard LMIs technique we immediately arrive at the following statement.

Theorem 6.5. *Pareto suboptimal output-feedback controller of the form (6.30) exists for the system (6.29) if and only if the LMIs*

$$\left(\begin{array}{c|c} N_1 & 0 \\ \hline 0 & I \end{array}\right)^T \left(\begin{array}{cccc|c} A^T X_{11} + X_{11} A & \star & \cdots & \cdots & \star \\ B_1^T X_{11} & -\alpha_1^2 \gamma^2 I & \star & \cdots & \star \\ \cdots & \cdots & \cdots & \cdots & \cdots \\ B_m^T X_{11} & 0 & \cdots & -\alpha_m^2 \gamma^2 I & \star \\ \hline C_z & D_{z,1} & \cdots & D_{z,m} & -I \end{array}\right) \left(\begin{array}{c|c} N_1 & 0 \\ \hline 0 & I \end{array}\right) < 0,$$

$$\left(\begin{array}{c|c} N_2 & 0 \\ \hline 0 & I \end{array}\right)^T \left(\begin{array}{cc|ccc} Y_{11} A^T + A Y_{11} & \star & \star & \cdots & \star \\ C_z Y_{11} & -I & \star & \cdots & \star \\ \hline B_1^T & D_{z,1}^T & -\alpha_1^2 \gamma^2 I & \star & \star \\ \cdots & \cdots & \cdots & \cdots & \cdots \\ B_m^T & D_{z,m}^T & 0 & \cdots & -\alpha_m^2 \gamma^2 I \end{array}\right) \left(\begin{array}{c|c} N_2 & 0 \\ \hline 0 & I \end{array}\right) < 0,$$

$$\begin{pmatrix} X_{11} & I \\ I & Y_{11} \end{pmatrix} \geq 0, \quad (B_{m+1} \ldots B_N)^T X_{11} (B_{m+1} \ldots B_N) < \gamma^2 \mathit{bdiag}\,(\alpha_{m+1}^2 I, \ldots, \alpha_N^2 I) \tag{6.34}$$

are feasible in $(n_x \times n_x)$*-matrices* $X_{11} = X_{11}^T > 0$, $Y_{11} = Y_{11}^T > 0$ *and* $\gamma^2 > 0$*, where columns of the matrices* N_1 *and* N_2 *form bases of kernels of matrices* $(C_y\, \mathcal{D}_y)$ *and* $(B_u^T\, D_u^T)$*, respectively.*

To find the gain matrix Θ_α^* one should insert the matrix X of the structure

$$X = \begin{pmatrix} X_{11} & X_{12} \\ X_{12}^{\mathrm{T}} & X_{22} \end{pmatrix},$$

where (X_{11}, Y_{11}) is a solution of LMIs (6.34) for the minimal value of γ^2 and $X_{12} = X_{22} = X_{11} - Y_{11}^{-1}$, into the first inequality (6.10) for $\mathcal{A} = \mathcal{A}_c$, $\mathcal{B}_v = \mathcal{B}_{cv}$, $\mathcal{C} = \mathcal{C}_c$, $\mathcal{D} = \mathcal{D}_c$ and solve this LMI with respect to Θ. The relative losses of these controllers are computed by the formula (6.20), where $v_1^*(t), \ldots, v_m^*(t), \hat{d}_{m+1}^*, \ldots, \hat{d}_N^*$ form the worst-case disturbance with respect to the performance measure (6.5) for the closed-loop system.

6.4. Filters

Consider the multi-objective filtering problem for a system described by the equations

$$\begin{aligned} \dot{x} &= Ax + \sum_{i=1}^{m} B_i w_i + \sum_{j=m+1}^{N} B_j w_j, \quad x(0) = 0, \\ y &= C_y x + \sum_{i=1}^{m} D_{y,i} w_i, \\ z &= C_z x + \sum_{i=1}^{m} D_{z,i} w_i, \end{aligned} \tag{6.35}$$

where $x \in R^{n_x}$ is the state, $y \in R^{n_y}$ is the measured output, $z \in R^{n_z}$ is the objective output, $w_i(t) \in L_2$, $i = 1, \dots, m$ and $w_j(t) = d_j \delta(t)$, $j = m+1, \dots, N$ are disturbances. The filter is assumed to be described by the equations

$$\begin{aligned} \dot{x}_f &= Ax_f + \Theta(y - C_y x_f), \quad x_f(0) = 0 \\ z_f &= C_z x_f. \end{aligned} \tag{6.36}$$

Then filtering errors $e(t) = x(t) - x_f(t)$ and $z(t) = \hat{z}(t) - z_f(t)$ are described by the equations of the form (6.3) with matrices given in (6.4), where

$$\begin{aligned} A(\Theta) &= A - \Theta C_y, \quad C(\Theta) = C_z, \\ B_i(\Theta) &= B_i - \Theta D_{y,i}, \; D_i(\Theta) = D_{z,i}, \quad i = 1, \dots, m, \\ B_j(\Theta) &= B_j, \quad j = m+1, \dots, N. \end{aligned} \tag{6.37}$$

By substituting the corresponding matrices into LMIs (6.10) and introducing the new variable $Z = X\Theta$, it is easy to check that the Γ_α-optimal filter is characterized as follows.

Theorem 6.6. *The gain matrices of Pareto suboptimal filters in the multi-objective filtering problem (6.16) for the system (6.35) are computed as* $\Theta_\alpha^* = X_*^{-1} Z_*$, *where parameters* $\alpha = (\alpha_1, \dots, \alpha_N)$ *belong to the* N*-dimensional simplex and* (X_*, Z_*, γ_*^2) *is a solution to the problem* $\inf \gamma^2 = \gamma_*^2$ *subject to constraints*

$$\begin{pmatrix} A^TX + XA - C_y^TZ^T - ZC_y & \star & \dots & \dots & \star \\ B_1^TX - D_{y,1}^TZ^T & -\alpha_1^2\gamma^2 I & \star & \dots & \star \\ \dots & \dots & \dots & \dots & \dots \\ B_m^TX - D_{y,m}^TZ^T & 0 & \dots & -\alpha_m^2\gamma^2 I & \star \\ C_z & D_{z,1} & \dots & D_{z,m} & -I \end{pmatrix} < 0, \tag{6.38}$$

$$(B_{m+1}\dots B_N)^T X (B_{m+1}\dots B_N) < \gamma^2 bdiag\,(\alpha_{m+1}^2 I, \dots, \alpha_N^2 I).$$

The relative losses of these filters compared with Pareto optimal ones are also computed by the formula (6.20), where the system matrices are given in (6.37).

6.5. State-Feedback Controllers Versus Coalitions of Disturbances

Given decomposition of the disturbance inputs $w_1, \dots, w_N$ on coalitions, Pareto suboptimal solutions in the corresponding multi-objective problem minimize the suboptimal cost $\hat{\Gamma}_\alpha(\Theta)$ for the system described by the equations (6.3), (6.4) with α_k replaced by $\alpha_{f(k)}\rho_{f(k)}$, $k = 1, \dots, N$. This means that Pareto suboptimal solutions in the control and filtering problems under coalitions of disturbances are found in similar way in view of the above substitutions. We present here, for example, the synthesis of the state-feedback controllers.

Theorem 6.7. *The gain matrix of the Pareto suboptimal state-feedback controller in the problem (5.2) with coalitions of disturbances defined by the function $l = f(k)$, $k = 1, \dots, N$ is computed as $\Theta_\alpha^* = ZY^{-1}$, where (Y, Z) is a solution of the problem* $\inf \gamma^2$ *subject to constraints*

$$\begin{pmatrix} AY + YA^T + B_uZ + Z^TB_u^T & \star & \dots & \dots & \star \\ B_1^T & -\alpha_{f(1)}^2\rho_1^2\gamma^2 I & \star & \dots & \star \\ \dots & \dots & \dots & \dots & \dots \\ B_m^T & 0 & \dots & -\alpha_{f(m)}^2\rho_m^2\gamma^2 I & \star \\ C_zY + D_uZ & D_{z,1} & \dots & D_{z,m} & -I \end{pmatrix} < 0,$$

$$\begin{pmatrix} Y & \star & \dots & \star \\ B_{m+1}^T & \alpha_{f(m+1)}^2\rho_{m+1}^2\gamma^2 I & \star & \star \\ \dots & \dots & \dots & \dots \\ B_N^T & 0 & \dots & \alpha_{f(N)}^2\rho_N^2\gamma^2 I \end{pmatrix} > 0. \tag{6.39}$$

7. Pareto Suboptimal Solutions for H_∞ and/or γ_0 Criteria in Discrete-time Systems

7.1. H_∞/γ_0-Norm

Consider an internally stable linear time-invariant system described by the equations

$$\begin{aligned} x(t+1) &= A(\Theta)x(t) + \sum_{i=1}^{m} B_i(\Theta)w_i(t) + \sum_{j=m+1}^{N} B_j(\Theta)w_j(t), \quad x(0)=0, \\ z(t) &= C(\Theta)x(t) + \sum_{i=1}^{m} D_i(\Theta)w_i(t) + \sum_{j=m+1}^{N} D_j(\Theta)w_j(t), \end{aligned} \tag{7.1}$$

where $x \in R^{n_x}$ is the state, $z \in R^{n_z}$ is the objective output. The system has disturbance signals of two types one of which $w_i \in R^{n_i}$, $w_i(0) = 0$, $i = 1,\dots,m$ are l_2-norm bounded signals and others $w_j \in R^{n_j}$, $j = m+1,\dots,N$ are signals of the form $w_j = d_j\delta_d(t)$, where $\delta_d(0) = 1$, $\delta_d(t) = 0$ for $t \geq 1$, d_j, $j = m+1,\dots,N$ are unknown vectors. The suboptimal multi-objective cost defined in (4.3) for this system takes the form

$$\Gamma_\alpha(\Theta) = \sup_{\substack{w_i \in l_2, i \in [1,m], \\ d_j \in R^{n_j}, j \in [m+1,N]}} \frac{\|z\|}{(\sum_{i=1}^{m} \alpha_i^2 \|w_i\|^2 + \sum_{j=m+1}^{N} \alpha_j^2 |d_j|^2)^{1/2}}. \tag{7.2}$$

Let us rewrite the equations (7.1) in the following form:

$$\begin{aligned} x(t+1) &= \mathcal{A}x(t) + \mathcal{B}_v v(t) + \mathcal{B}_\zeta \zeta(t), \\ z(t) &= \mathcal{C}x(t) + \mathcal{D}_v v(t) + \mathcal{D}_\zeta \zeta(t), \end{aligned} \tag{7.3}$$

where $v = \text{col}(v_1,\dots,v_m) \in l_2$, $\zeta = \hat{d}\delta_d(t)$, $\hat{d} = \text{col}(\hat{d}_{m+1},\dots,\hat{d}_N) \in R^{\hat{n}}$, $\hat{n} = \sum_{j=m+1}^{N} n_j$,

$$\begin{aligned} \mathcal{B}_v(\Theta) &= (\alpha_1^{-1}\mathcal{B}_1(\Theta),\dots,\alpha_m^{-1}\mathcal{B}_m(\Theta)), \quad \mathcal{B}_\zeta(\Theta) = (\alpha_{m+1}^{-1}\mathcal{B}_{m+1}(\Theta),\dots,\alpha_N^{-1}\mathcal{B}_N(\Theta)), \\ \mathcal{D}_v(\Theta) &= (\alpha_1^{-1}\mathcal{D}_1(\Theta),\dots,\alpha_m^{-1}\mathcal{D}_m(\Theta)). \quad \mathcal{D}_\zeta(\Theta) = (\alpha_{m+1}^{-1}\mathcal{D}_{m+1}(\Theta),\dots,\alpha_N^{-1}\mathcal{D}_N(\Theta)). \end{aligned} \tag{7.4}$$

It is clear that for $v_i = \alpha_i w_i$, $i = 1,\dots,m$ and $\hat{d}_j = \alpha_j d_j$, $j = m+1,\dots,N$, the equations (7.1) and (7.3) are the same, whereas the performance measure (7.2) takes the form

$$\gamma_{\infty,0} = \sup_{v \in l_2, \hat{d} \in R^{\hat{n}}} \frac{\|z\|}{(\|v\|^2 + |\hat{d}|^2)^{1/2}}. \tag{7.5}$$

This is the worst-case l_2-norm of the objective output over all admissible signals acting on two disturbances inputs or, in other words, the H_∞/γ_0 norm of the system. The following statements concerning the calculation of both this performance measure and the worst-case disturbances can be proved similarly to the above continuous-time case.

Theorem 7.1. *The H_∞/γ_0 norm $\gamma_{\infty,0}$ is less than γ in the system (7.3) if and only if*

$$\lambda_{max}(\mathcal{B}_\zeta^T X_\gamma \mathcal{B}_\zeta + \mathcal{D}_\zeta^T \mathcal{D}_\zeta) < \gamma^2, \tag{7.6}$$

where $X_\gamma = X_\gamma^T \geq 0$ is the solution to the Riccati equation

$$\mathcal{A}^T X \mathcal{A} - X + \mathcal{C}^T \mathcal{C} + (\mathcal{B}_v^T X \mathcal{A} + \mathcal{D}_v^T \mathcal{C})^T (\gamma^2 I - \mathcal{B}_v^T X \mathcal{B}_v - \mathcal{D}_v^T \mathcal{D}_v)^{-1} (\mathcal{B}_v^T X \mathcal{A} + \mathcal{D}_v^T \mathcal{C}) = 0 \tag{7.7}$$

such that $\gamma^2 I - \mathcal{B}_v^T X_\gamma \mathcal{B}_v - \mathcal{D}_v^T \mathcal{D}_v > 0$ and

$$\mathcal{A}_v = \mathcal{A} + \mathcal{B}_v(\gamma^2 I - \mathcal{B}_v^T X_\gamma \mathcal{B}_v - \mathcal{D}_v^T \mathcal{D}_v)^{-1} (\mathcal{B}_v^T X_\gamma \mathcal{A} + \mathcal{D}_v^T \mathcal{C}) \tag{7.8}$$

is a Schur stable matrix.

Corollary 7.1. *The H_∞/γ_0 norm for the system (7.3) can be computed in solving the problem $\inf\limits_{(7.9)} \gamma^2$ subject to LMIs*

$$\begin{pmatrix} -X & \star & \star & \star \\ \mathcal{A}^T X & -X & \star & \star \\ \mathcal{B}_v^T X & 0 & -\gamma^2 I & \star \\ 0 & \mathcal{C} & \mathcal{D}_v & -I \end{pmatrix} \leq 0, \quad \begin{pmatrix} -X & \star & \star \\ \mathcal{B}_\zeta^T X & -\gamma^2 I & \star \\ 0 & \mathcal{D}_\zeta & -I \end{pmatrix} \leq 0. \tag{7.9}$$

Theorem 7.2. *Let the following inequality holds*

$$\lambda_{max}(\mathcal{B}_\zeta^T X_\infty \mathcal{B}_\zeta + \mathcal{D}_\zeta^T \mathcal{D}_\zeta) > \gamma_\infty^2, \tag{7.10}$$

where X_∞ is the solution to the first inequality in (7.9) at the minimal value of γ^2 equal to γ_∞^2. Then supremum in (7.5) is actually achieved and eaual to γ_α under the worst-case disturbance

$$v^*(t) = L_* x_*(t), \quad L_* = (\gamma_\alpha^2 I - \mathcal{B}_v^T X_* \mathcal{B}_v - \mathcal{D}_v^T \mathcal{D}_v)^{-1} (\mathcal{B}_v^T X_* \mathcal{A} + \mathcal{D}_v^T \mathcal{C}) \tag{7.11}$$

and $\hat{d}^* = e_{max}(\mathcal{B}_\zeta^T X_* \mathcal{B}_\zeta + \mathcal{D}_\zeta^T \mathcal{D}_\zeta)$, where $X_* = X_*^T > 0$ is the stabilizing solution of the Riccati equation (7.7) for $\gamma = \gamma_\alpha$, $x_*(t)$ is the solution of the system

$$x(t+1) = \mathcal{A}_v x(t) + \mathcal{B}_\zeta \hat{d}^* \delta(t), \quad x(0) = 0, \tag{7.12}$$

$\mathcal{A}_v$ is the Schur stable matrix given in (7.8) for $\gamma = \gamma_\alpha$ and $X_\gamma = X_*$, e_{max} is the normalized eigenvector associated with the maximum eigenvalue γ_α^2 of the matrix $\mathcal{B}_\zeta^T X_* \mathcal{B}_\zeta + \mathcal{D}_\zeta^T \mathcal{D}_\zeta$.

Remark 7.1. *When the inequality (7.10) does not hold, i.e.*

$$\lambda_{max}(\mathcal{B}_\zeta^T X_\infty \mathcal{B}_\zeta + \mathcal{D}_\zeta^T \mathcal{D}_\zeta) \leq \gamma_\infty^2, \tag{7.13}$$

and, hence, $\gamma_{\infty,0} = \|H_v\|_\infty$, the sequence $v_\varepsilon(t) = L_* x_*(t)$ with L_* given in (7.11) for $\gamma_\alpha = \gamma_\varepsilon = \|H_v\|_\infty + \varepsilon$ and $X_* = X_\varepsilon$, as well as $\zeta(t) = \hat{d}\delta(t)$ with an arbitrary $\hat{d}$ approach the worst-case disturbance in (7.5) as $\varepsilon \to 0$. Here X_ε is the stabilizing solution of the Riccati equation (7.7) for $\gamma = \gamma_\varepsilon$, $x_\varepsilon(t)$ is the solution of the system (7.12), where $\mathcal{A}_v$ is a Schur stable matrix given in (7.8) for $\gamma = \gamma_\varepsilon$ and $X_\gamma = X_\varepsilon$. Indeed, since

$$\|z\|^2 - \gamma_\varepsilon^2 \|v_\varepsilon\|^2 = \hat{d}^T(\mathcal{B}_\zeta^T X_\varepsilon \mathcal{B}_\zeta + \mathcal{D}_\zeta^T \mathcal{D}_\zeta)\hat{d}$$

and some eigenvalues of the matrix $\mathcal{A}_v$ tend to the unit circle as $\varepsilon \to 0$, we conclude that $\|v_\varepsilon\| \to \infty$ and, hence, for any $\hat{d}$

$$\lim_{\varepsilon \to 0} \frac{\|z\|^2}{\|v_\varepsilon\|^2 + |\hat{d}|^2} = \lim_{\varepsilon \to 0} \frac{\gamma_\varepsilon^2 \|v_\varepsilon\|^2 + \hat{d}^T(\mathcal{B}_\zeta^T X_\varepsilon \mathcal{B}_\zeta + \mathcal{D}_\zeta^T \mathcal{D}_\zeta)\hat{d}}{\|v_\varepsilon\|^2 + |\hat{d}|^2} = \|H_v\|_\infty^2.$$

Now we will apply the above characterizations of the suboptimal multi-objective cost and the worst-case disturbance to compute Γ_α-optimal solutions as Pareto suboptimal ones in the multi-objective filtering and control problems and evaluate their relative losses compared with Pareto optimal solutions.

7.2. Filters

Consider the multi-objective filtering problem for a system described by the equations

$$\begin{aligned}
\hat{x}(t+1) &= A\hat{x}(t) + \sum_{i=1}^{m} B_i w_i(t) + \sum_{j=m+1}^{N} B_j w_j(t), \quad \hat{x}(0) = 0, \\
\hat{y}(t) &= C_y \hat{x}(t) + \sum_{i=1}^{m} D_{y,i} w_i(t) + \sum_{j=m+1}^{N} D_{y,j} w_j(t), \\
\hat{z}(t) &= C_z \hat{x}(t) + \sum_{i=1}^{m} D_{z,i} w_i(t) + \sum_{j=m+1}^{N} D_{z,j} w_j(t)
\end{aligned} \tag{7.14}$$

where $\hat{x}(t) \in R^{n_x}$ is the state, $\hat{y}(t) \in R^{n_y}$ is the measured output, $\hat{z}(t) \in R^{n_z}$ is the objective output, $w_i(t) \in l_2$, $i = 1, \dots, m$, $w_j(t) = d\delta_d(t)$, $j = m+1, \dots, N$. The filter is assumed to be described by the equations

$$x_f(t+1) = Ax_f(t) + \Theta[\hat{y}(t) - C_y x_f(t)], \quad x_f(0) = 0$$
$$z_f(t) = C_z x_f(t). \tag{7.15}$$

Then filtering errors $x(t) = \hat{x}(t) - x_f(t)$ and $z(t) = \hat{z}(t) - z_f(t)$ are described by the equations of the form (7.3), where

$$\mathcal{A}(\Theta) = A - \Theta C_y,\ \mathcal{C}(\Theta) = C_z,\ \mathcal{B}_k(\Theta) = B_k - \Theta D_{y,k},\ \mathcal{D}_k(\Theta) = D_{z,k},\ k = 1, \dots, N. \tag{7.16}$$

By substituting the corresponding matrices given in (7.4), (7.16) into LMIs (7.9) and introducing the new variable $Z = X\Theta$, it is easy to check that the Γ_α-optimal filter is characterized as follows.

Theorem 7.3. *The gain matrices of Pareto suboptimal filters in the multi-objective filtering problem (6.16) for the system (7.14) are computed as* $\Theta_\alpha^* = X_*^{-1} Z_*$, *where parameters* $\alpha = (\alpha_1, \dots, \alpha_N)$ *belong to* N*-dimensional simplex and* (X_*, Z_*, γ_*^2) *is the solution to the problem* $\inf_{(7.17)} \gamma^2 = \gamma_*^2$ *subject to constraints*

$$\begin{pmatrix} -X & \star & \dots & \dots & \dots & \star \\ A^T X - C_y^T Z^T & -X & \star & \dots & \dots & \star \\ B_1^T X - D_{y,1}^T Z^T & 0 & -\alpha_1^2\gamma^2 I & \star & \dots & \star \\ \dots & \dots & \dots & \dots & \dots & \dots \\ B_m^T X - D_{y,m}^T Z^T & 0 & \dots & 0 & -\alpha_m^2\gamma^2 I & \star \\ 0 & C_z & D_{z,1} & \dots & D_{z,m} & -I \end{pmatrix} \le 0,$$

$$\begin{pmatrix} -X & \star & \dots & \dots & \star \\ B_{m+1}^T X - D_{y,m+1}^T Z^T & -\alpha_{m+1}^2\gamma^2 I & \star & \dots & \star \\ \dots & \dots & \dots & \dots & \dots \\ B_N^T X - D_{y,N}^T Z^T & 0 & \dots & -\alpha_N^2\gamma^2 I & \star \\ 0 & D_{z,m+1} & \dots & D_{z,N} & -I \end{pmatrix} \le 0. \tag{7.17}$$

According to Theorem 4.2 the relative losses of the Γ_α-optimal filter do not exceed $1 - \mu(\alpha)$, where

$$\mu(\alpha) = \frac{(\sum_{i=1}^{m} \alpha_i^2 \|w_i^*\|^2 + \sum_{j=m+1}^{N} \alpha_j^2 |d_j^*|^2)^{1/2}}{\sum_{i=1}^{m} \alpha_i \|w_i^*\| + \sum_{j=m+1}^{N} \alpha_j |d_j^*|}. \tag{7.18}$$

These relative losses can be equivalently expressed in terms of the worst-case disturbances with respect to performance measure (7.5) for the system (7.3) as follows

$$\mu(\alpha) = \frac{\sqrt{\|v^*\|^2 + 1}}{\sum_{i=1}^{m} \|v_i^*\| + \sum_{j=m+1}^{N} |\hat{d}_j^*|}, \tag{7.19}$$

where $v^* = \mathrm{col}(v_1^*, \dots, v_m^*)$, $v^*(t) = L_* x_*(t)$, $v_i^* = L_i x_*(t)$, $L_* = (L_1, \dots, L_m)$, $\hat{d}^* = \mathrm{col}(d_{m+1}^*, \dots, d_N^*)$, $\hat{d}^* = e_{max}(\mathcal{B}_\zeta^T X_* \mathcal{B}_\zeta + \mathcal{D}_\zeta^T \mathcal{D}_\zeta)$, $(|\hat{d}^*| = 1)$ are determined in Theorem 7.2 provided that A is replaced by $A - \Theta_\alpha^* C_y$. Note that $\|v_i^*(t)\|^2 = \hat{d}^{*T} \mathcal{B}_\zeta{}^T X_0 \mathcal{B}_\zeta \hat{d}^*$, where $X_0 = X_0^T \geq 0$ is a solution to the Lyapunov equation

$$\mathcal{A}_v^T X \mathcal{A}_v - X + L_i^T L_i = 0 \tag{7.20}$$

in which $\mathcal{A}_v = A - \Theta_\alpha^* C_y + \mathcal{B}_v L_*$ is a Schur stable matrix.

In the case when all criteria in (6.16) are H_∞ norms, the suboptimal multi-objective cost takes the form

$$\Gamma_\alpha(\Theta) = \sup_{w_i \in l_2, i \in [1,N]} \frac{\|z\|}{(\sum_{i=1}^{N} \alpha_i^2 \|w_i\|^2)^{1/2}} = \sup_{v \in l_2, \|v\| \neq 0} \frac{\|z\|}{\|v\|}. \tag{7.21}$$

This is the H_∞ norm of the transfer matrix $H_{zv}(s)$ from v to z of the system (7.3), which is characterized by the first inequality in (7.9). The gain matrices of the Pareto suboptimal filters are computed as $\Theta_\alpha^* = Z_* Y_*^{-1}$, where the pair of matrices (Y_*, Z_*) is a solution to the problem $\inf \gamma^2$ subject to the first inequality in (7.17) with $m = N$. According to (7.18), the relative losses do not exceed

$1-\mu(\alpha)$, where

$$\mu(\alpha)=\lim_{\varepsilon\to 0}\frac{(\sum\limits_{i=1}^{N}\alpha_i^2\|w_{i\varepsilon}^*\|^2)^{1/2}}{\sum\limits_{i=1}^{N}\alpha_i\|w_{i\varepsilon}^*\|}=\lim_{\varepsilon\to 0}\frac{\|v_\varepsilon\|}{\sum\limits_{i=1}^{N}\|v_{i\varepsilon}\|} \tag{7.22}$$

and $v_\varepsilon(t)$ approaches the worst-case disturbance in (7.21). To compute $\|v_{i\varepsilon}\|$ we artificially introduce the signal $\zeta(t)=\hat{d}\delta_d(t)$ in such a way to obtain the system in the form (7.3). Choose $\mathcal{B}_\zeta=\nu I$, $\mathcal{D}_\zeta=0$ and parameter ν such that the corresponding inequality (7.13) would be valid (for instance, $\nu^2\lambda_{max}(X_\infty)\leq\|H_v\|_\infty^2$). Then taking into account Remark 7.1, we set $v_\varepsilon=\mathrm{col}(v_{1\varepsilon},\ldots,v_{N\varepsilon})=L_*x_\varepsilon(t)$, $v_{i\varepsilon}=L_{i*}x_\varepsilon(t)$, $L_\varepsilon=(L_{1*},\ldots,L_{N*})$ and compute $\|v_{i\varepsilon}(t)\|^2=\hat{d}^{*T}\mathcal{B}_\zeta{}^TX_0\mathcal{B}_\zeta\hat{d}^*$, where $X_0=X_0^T\geq 0$ is a solution to the Lyapunov equation (7.20).

In the case when all criteria in (6.16) are γ_0 norms, the suboptimal multi-objective cost takes the form

$$\Gamma_\alpha(\Theta)=\sup_{d_j,j\in[1,N]}\frac{\|z\|}{(\sum\limits_{j=1}^{N}\alpha_j^2|d_j|^2)^{1/2}}=\sup_{\hat{d}\neq 0}\frac{\|z\|}{|\hat{d}|}. \tag{7.23}$$

This is the γ_0 norm of the corresponding transfer matrix and the gain matrices of Pareto suboptimal filters are computed in solving the LMIs (7.17), where the block rows and columns with numbers $3,\ldots,N+2$ should be excluded in the first inequality and $m=0$ in the second inequality. The relative losses of Pareto suboptimal filters do not exceed $1-\mu(\alpha)$, where

$$\mu(\alpha)=\frac{(\sum\limits_{j=1}^{N}\alpha_j^2|d_j^*\|^2)^{1/2}}{\sum\limits_{j=1}^{N}\alpha_j|d_j^*|}=\frac{1}{\sum\limits_{j=1}^{N}|\hat{d}_j^*|}. \tag{7.24}$$

The worst-case disturbance $\zeta^*(t)=\hat{d}^*\delta_d(t)$, where $\hat{d}^*$ is the normalized eigenvector associated with the maximum eigenvalue of the matrix $\mathcal{B}_\zeta^TX_0\mathcal{B}_\zeta+\mathcal{D}_\zeta^T\mathcal{D}_\zeta$ (see Theorem 7.2).

7.3. State-Feedback Controllers

Consider the multi-objective control problem for the system

$$\begin{aligned} x(t+1) &= Ax(t) + \sum_{i=1}^{m} B_i w_i(t) + \sum_{j=m+1}^{N} B_j w_j(t) + B_u u, \quad x(0) = x_0, \\ z(t) &= C_z x(t) + \sum_{i=1}^{m} D_i w_i(t) + \sum_{j=m+1}^{N} D_j w_j(t) + D_u u \end{aligned} \tag{7.25}$$

in the class of linear state-feedbacks $u = \Theta x$. The closed-loop system is described by the equations (7.3), where

$$\mathcal{A}(\Theta) = A + B_u\Theta, \quad \mathcal{C} = C_z + D_u\Theta, \quad \mathcal{B}_k(\Theta) = B_k, \quad \mathcal{D}_k(\Theta) = D_k,\ k = 1, \dots, N. \tag{7.26}$$

By substituting the matrix of the closed-loop system into (7.9), multiplying the first and second inequalities by diag (X^{-1}, X^{-1}, I, I) and diag (X^{-1}, I, I), respectively, from the left and from the right, introducing the new variables $Y = X^{-1}$ and $Z = \Theta Y$, it is easy to check the that LMIs (7.9) result in the following characterization of the Γ_α-optimal controller.

Theorem 7.1. *The gain matrices of the Pareto suboptimal controllers in the multi-objective control problem (6.16) for the system (7.25) are calculated as $\Theta^*_\alpha = Z_* Y_*^{-1}$, where parameters $\alpha = (\alpha_1, \dots, \alpha_N)$ belong to N-dimensional simplex and (Y_*, Z_*, γ_*^2) is the solution to the problem $\inf\limits_{(7.27)} \gamma^2 = \gamma_*^2$ subject to constraints*

$$\begin{pmatrix} -Y & \star & \dots & \dots & \dots & \star \\ YA^T + Z^T B_u^T & -Y & \star & \dots & \dots & \star \\ B_1^T & 0 & -\alpha_1^2\gamma^2 I & \star & \dots & \star \\ \dots & \dots & \dots & \dots & \dots & \dots \\ B_m^T & 0 & \dots & 0 & -\alpha_m^2\gamma^2 I & \star \\ 0 & C_z Y + D_u Z & D_1 & \dots & D_m & -I \end{pmatrix} \le 0,$$

$$\begin{pmatrix} -Y & \star & \dots & \dots & \star \\ B_{m+1}^T & -\alpha_{m+1}^2\gamma^2 I & \star & \dots & \star \\ \dots & \dots & \dots & \dots & \dots \\ B_N^T & 0 & \dots & -\alpha_N^2\gamma^2 I & \star \\ 0 & D_{m+1} & \dots & D_N & -I \end{pmatrix} \le 0. \tag{7.27}$$

The relative losses of these controllers are computed in a similar manner as for the filters. Note that if all the criteria in (6.16) are γ_0 norms, the gain matrices of the γ_0-optimal controllers do not depend on the parameters α_k, $k = 1, \ldots, N$. Indeed, it is easy to see by completing the square in the corresponding Lyapunov equation when $\mathcal{A}(\Theta) = A + B_u\Theta$, $\mathcal{C}(\Theta) = C_z + D_u\Theta$. In this case there is the single Pareto optimal controller coinciding with the linear quadratic optimal controller fo which $\Theta_* = -(B_u^T X_* B_u + D_u^T D_u)^{-1}(B_u^T X_* A + D_u^T C)$, where $X_* = X_*^T > 0$ is the stabilizing solution to the Riccati equation

$$A^T XA - X + C_z^T C_z - (B_u^T XA + D_u^T C_z)^T (B_u^T XB_u + D_u^T D_u)^{-1}(B_u^T XA + D_u^T C_z) = 0.$$

8. The Dual Multi-Objective Problem

Consider an internally stable system with N channels from a single disturbance input to N objective outputs described by the equations

$$\begin{aligned} \partial x &= A(\Theta)x + B_\zeta(\Theta)\zeta, \quad x(0) = 0, \\ z_k &= C_k(\Theta)x, \quad k = 1, \ldots, N, \end{aligned} \tag{8.1}$$

where $\zeta \in R^{n_\zeta}$ is the disturbance input and $z_k \in R^{n_k}$, $k = 1, \ldots, N$ are the objective outputs, Θ is the matrix of parameters. Let $H_k(s)$, $k = 1, \ldots, N$ be a transfer matrix of the system (8.1) from ζ to z_k. Associated to each channel from ζ to z_i, $i = 1, \ldots, m$ is defined the following criterion:

$$\tilde{J}_i(\Theta) = \sup_{\zeta(t) \in L_2, \|\zeta\| \neq 0} \frac{\|z_i\|}{\|\zeta\|} = \|H_i(\Theta)\|_\infty, \tag{8.2}$$

whereas associated to each channel from ζ to z_j, $j = m+1, \ldots, N$ is defined the following criterion:

$$\tilde{J}_j(\Theta) = \sup_{\zeta(t) \in L_2, \|\zeta\| \neq 0} \frac{\|z_j\|_\infty}{\|\zeta\|} = \|H_j(\Theta)\|_{g2}. \tag{8.3}$$

The problem under consideration is to characterize the Pareto optimal set, i.e.

$$\Theta_P = \arg\min_{\Theta}\{\|H_i(\Theta)\|_\infty,\ i = 1, \ldots, m;\ \|H_j(\Theta)\|_{g2},\ j = m+1, \ldots, N\}. \tag{8.4}$$

Consider the dual system with the corresponding N channels from w_k, $k = 1, \ldots, N$ to z described by the equations

$$\begin{aligned} \partial x &= A^T(\Theta)x + \sum_{i=1}^{m} C_i^T(\Theta)w_i + \sum_{j=m+1}^{N} C_j^T(\Theta)w_j, \\ z &= B_\zeta^T(\Theta)x, \quad x(0) = 0, \end{aligned} \tag{8.5}$$

Since $\|H_i(\Theta)\|_\infty = \|H_i^T(\Theta)\|_\infty$ and $\|H_j(\Theta)\|_{g2} = \gamma_0(H_j^T(\Theta))$, the Pareto suboptimal solutions in the multi-objective problem under consideration for a system with a single disturbance input and multiple objective outputs can be computed by means of solving the corresponding multi-objective problem for the dual system with the multiple inputs and the single output.

In particular cases, when the criteria in (8.4) are all H_∞ norms or all H_{g2} norms, the suboptimal multi-objective costs are the H_∞ or γ_0 norm, respectively, of the system

$$\begin{aligned} &\partial x = A^T(\Theta)x + \mathcal{C}^T(\Theta)\zeta, \quad \mathcal{C}^T(\Theta) = (\alpha_1^{-1}C_1^T(\Theta), \ldots, \alpha_N^{-1}C_N^T(\Theta)), \\ &z = B_\zeta^T(\Theta)x \end{aligned} \tag{8.6}$$

with the single disturbance input and the single objective output. In turn, these are the H_∞ or H_{g2} norm, respectively, of the system

$$\partial x = A(\Theta)x + B_\zeta(\Theta)\zeta, \quad z = \begin{pmatrix} \beta_1 z_1 \\ \cdots \\ \beta_N z_N \end{pmatrix}, \tag{8.7}$$

where $\beta_k \geq 0$, $\sum_{k=1}^{N} \beta_k = 1$, and the objective output z is a linear combination of the objective outputs z_k of the initial system (8.1). The relative losses of the corresponding Pareto suboptimal solutions compared with the Pareto optimal ones do not exceed $1 - \sqrt{N}/N$.

9. Illustrative Examples

Consider the double-inverted pendulum

$$\begin{aligned} \ddot{\varphi}_1 &= 2\varphi_1 - \varphi_2 + u\,, \\ \ddot{\varphi}_2 &= -2\varphi_1 + 2\varphi_2\,, \end{aligned}$$

where φ_1 and φ_2 are the deflection angles of the pendulum links from the vertical positions, u is a control input. The initial conditions for pendulum links are assumed to be unknown. The goal of control is to provide tradeoff between optimal attenuation of the disturbances caused by external excitations and the transients caused by nonzero initial conditions.

At first we consider the state-feedback case. Rewrite this system in the form of (6.15), where

$$A = \begin{pmatrix} 0 & 0 & 1 & 0 \\ 0 & 0 & 0 & 1 \\ 2 & -1 & 0 & 0 \\ -2 & 2 & 0 & 0 \end{pmatrix}, \; B_1 = \begin{pmatrix} 0 \\ 0 \\ 0 \\ 1 \end{pmatrix}, \; B_2 = I_{4\times 4},$$

$$B_u = \begin{pmatrix} 0 \\ 0 \\ 1 \\ 0 \end{pmatrix}, \; C_z = \begin{pmatrix} 1 & 0 & 0 & 0 \\ 0 & 1 & 0 & 0 \\ 0 & 0 & 0 & 0 \end{pmatrix}, \; D_u = \begin{pmatrix} 0 \\ 0 \\ 1 \end{pmatrix},$$

$w_1 \in L_2$ and $w_2 = d\delta(t)$, $d = x(0)$ is an unknown vector. Consider the two-criteria problem

$$\Theta_P = \arg\min_{\Theta}\{J_k(\Theta),\, k = 1, 2\},$$

where

$$J_1(\Theta) = \sup_{\|w_1\|\neq 0} \frac{\|z\|}{\|w_1\|}, \quad J_2(\Theta) = \sup_{|d|\neq 0} \frac{\|z\|}{|d|}.$$

The gain matrices of controllers minimizing each of these criteria are as follows

$$\Theta_\infty = 10^{11}(1.75 \quad -2.10 \quad 0.43 \quad -1.50),$$
$$\Theta_0 = (-14.42 \quad 14.42 \quad -5.46 \quad 9.51)$$

and the minimal values of these criteria are $J_1(\Theta_\infty) = 3.46$ and $J_2(\Theta_0) = 12.07$. The points labeled by Θ_∞ and Θ_0 on the plane of criteria (J_2, J_1) correspond to the values of the criteria for the H_∞ and γ_0 optimal controllers, respectively (see Fig. 1). The Pareto optimal set is located in the shaded curvelinear triangle shown on Fig. 1. The boundaries of this domain consist of two right lines corresponding to the minimal values of the criteria and the curve line. The coordinates of curve points correspond to $(J_2(\Theta_\alpha^*), J_1(\Theta_\alpha^*))$, where Θ_α^* are

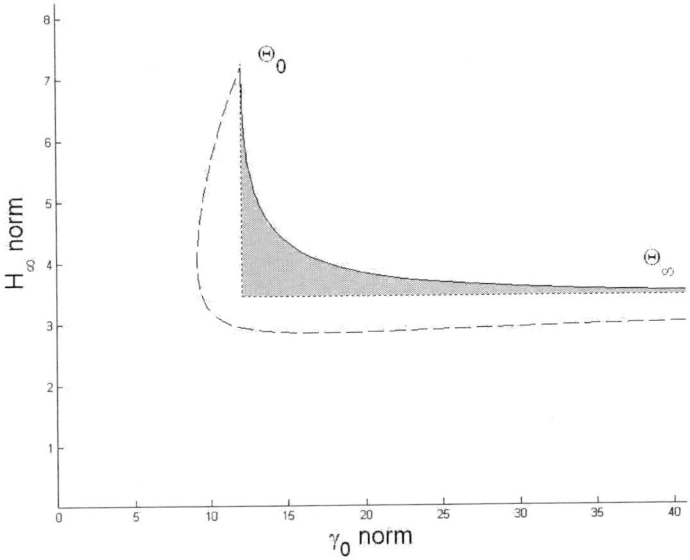

Figure 1. The estimation of the Pareto optimal set in the space of γ_0 and H_∞ criteria: the state-feedback case.

the H_∞/γ_0 optimal solutions to the performance measure

$$\Gamma_\alpha(\Theta) = \sup_{w_1 \in L_2, d \in R^4} \frac{\|z\|}{\sqrt{\alpha^2\|w_1\|^2 + (1-\alpha)^2|d|^2}}, \qquad \alpha \in (0,1).$$

The coordinates of the dotted curve on Fig. 1 correspond to $(\mu(\alpha)J_2(\Theta^*_\alpha)$, $\mu(\alpha)J_1(\Theta^*_\alpha))$, where $\mu(\alpha)$ is defined in (6.19). This curve could determine the lower boundary of a domain estimating the Pareto optimal set if it were located inside the shaded triangle. In the next examples we will see that this curve plays an important role in estimating the Pareto optimal set.

Now, we synthesize the output-feedback controller for the same plant with the measured output

$$y = \begin{pmatrix} 1 & 0 & 0 & 0 \\ 0 & 1 & 0 & 0 \end{pmatrix} x$$

and consider the same two-criteria problem. The matrices of the output-feedback controllers minimizing each of these criteria, Θ_∞ and Θ_0, are as follows

$$\Theta_\infty = 10^3 \begin{pmatrix} -16.6 & 4.0 & -0.55 & 2.1 & 14.3 & -1.1 \\ 0.26 & -0.5 & 0.049 & -0.18 & -0.058 & 0.24 \\ -9.7 & -1.9 & -3.9 & 14.4 & -6.1 & 22.5 \\ -0.33 & -19.1 & 1.5 & -5.7 & 6.6 & 10.93 \\ -4.9 & -3.1 & -3.7 & 13.9 & -10.21 & 22.8 \end{pmatrix},$$

$$\Theta_0 = \begin{pmatrix} -19.33 & 1.55 & -1.57 & 5.30 & 11.85 & 6.32 \\ 2.74 & -19.33 & 2.17 & -7.84 & 6.48 & 6.64 \\ 1.43 & -6.57 & -5.09 & 9.06 & -13.02 & 19.43 \\ 0.37 & -5.21 & 0.57 & -2.37 & 0.094 & 3.81 \\ 2.62 & -2.38 & -5.68 & 11.55 & -18.77 & 19.81 \end{pmatrix},$$

and the minimal values of these criteria are $J_1(\Theta_\infty) = 3.81$ and $J_2(\Theta_0) = 12.97$. The Pareto optimal set is located in the shaded domain shown on Fig. 2.

Finally, we consider two coalitions of exogenous disturbances, one of them $\eta_1 = \{w_1, w_3\}$ and another coalition $\eta_2 = \{w_2, w_4\}$, where w_1 and w_2 are the same as before, while $w_3 \in L_2$ and $w_4 \in L_2$ are disturbances in the measured output

$$y = \begin{pmatrix} 1 & 0 & 0 & 0 \\ 0 & 1 & 0 & 0 \end{pmatrix} x + \begin{pmatrix} 1 \\ 1 \end{pmatrix} w_3 + \begin{pmatrix} 1 \\ 1 \end{pmatrix} w_4.$$

Consider two-criteria problem

$$\Theta_P = \arg\min_{\Theta}\{\hat{J}_k(\Theta),\, k = 1, 2\},$$

where

$$\hat{J}_1(\Theta) = \sup_{\|\eta_1\|\neq 0} \frac{\|z\|}{\|\eta_1\|} = \sup_{\|\eta_1\|\neq 0} \frac{\|z\|}{\sqrt{\rho_1^2\|w_1\|^2 + \rho_3^2\|w_3\|^2}},$$

$$\hat{J}_2(\Theta) = \sup_{\|\eta_2\|\neq 0} \frac{\|z\|}{\|\eta_2\|} = \sup_{\|\eta_2\|\neq 0} \frac{\|z\|}{\sqrt{\rho_2^2|d|^2 + \rho_4^2\|w_4\|^2}}.$$

The optimal values of these criteria for $\rho_1 = \rho_2 = \rho_4 = 1$ and $\rho_3 = 3$ are $\hat{J}_1(\Theta_{\infty,\infty}) = 4.92$ and $\hat{J}_2(\Theta_{\infty,0}) = 13.01$. The points labeled by $\Theta_{\infty,\infty}$ and $\Theta_{\infty,0}$ on the plane of criteria $(\hat{J}_2, \hat{J}_1)$ correspond to the values of the criteria for

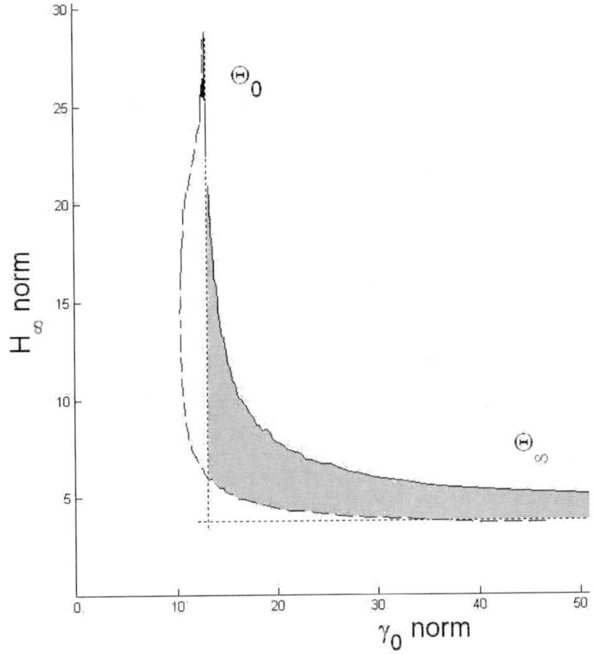

Figure 2. The estimation of the Pareto optimal set in the space of γ_0 and H_∞ criteria: the output-feedback case.

the optimal controllers (see Fig. 3). The matrices Θ_α^* of the Pareto suboptimal controllers are found as the optimal solutions to the performance measure

$$\hat{\Gamma}_\alpha(\Theta) = \sup_{\eta_1,\eta_2} \frac{\|z\|}{\sqrt{\alpha^2\|\eta_1\|^2 + (1-\alpha)^2\|\eta_2\|^2}}, \qquad \alpha \in (0,1).$$

The Pareto optimal set is located in the shaded domain shown on Fig. 3.

Conclusion

An universal framework for analyzing disturbance attenuation levels of linear systems under multiple potentially possible disturbance inputs and synthesizing Pareto suboptimal solutions in the multi-objective control and filtering problems is considered in this chapter. The relative losses of the optimal solutions with respect to a multi-objective cost compared to the Pareto optimal solutions are

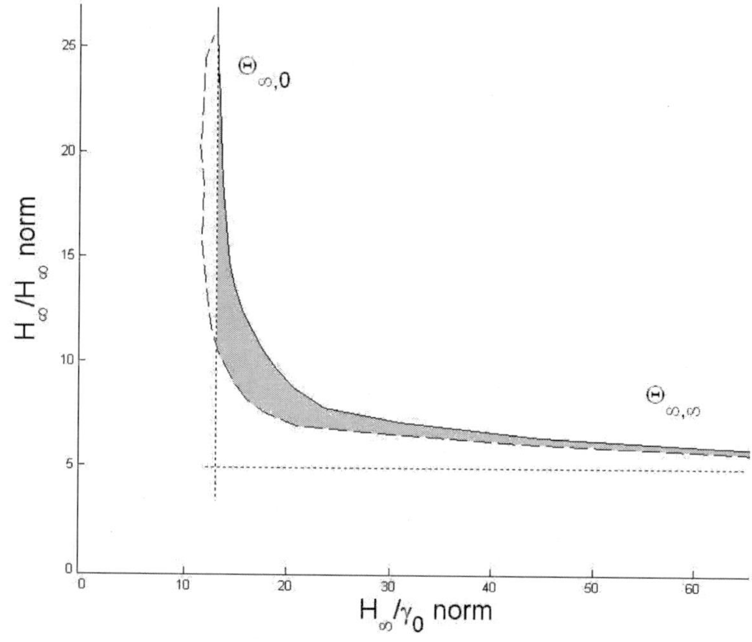

Figure 3. The estimation of the Pareto optimal set in the case of coalitions: H_∞/γ_0 versus H_∞/H_∞.

established to be less than or equal to $1 - \sqrt{N}/N$, where N is the number of disturbance inputs. The approach also allows to analyze various situations when disturbances acting on different inputs are combined into coalitions and synthesize the corresponding Pareto suboptimal solutions.

It is shown that for classes of L_2-bounded and impulsive disturbances with the H_∞ and/or γ_0 norms as the disturbance attenuation levels, the multi-objective cost is the H_∞/γ_0 norm of the system. The Pareto suboptimal H_∞/γ_0 controllers and filters are synthesized in terms of LMIs and their relative losses are evaluated. The analogous problems can be also solved for various classes of disturbances with both the H_∞ and γ_0 norms as the disturbance attenuation levels as well as other disturbance attenuation levels.

References

[1] Bernstein, D.S. and Haddad, W.M. (1989). LQG control with an H_∞ performance bound: a Riccati equation approach. *IEEE Trans. Autom. Control*, vol. 34, no. 3, pp. 293–305.

[2] Khargonekar, P.P. and Rotea, M.A. (1991). Mixed H_2/H_∞ control: a convex optimization approach. *IEEE Trans. Autom. Control*, vol. 36, no. 7, pp. 824–831.

[3] Zhou, K., Glover, K., Bodenheimer, B. and Doyle, J. (1994). Mixed H_2 and H_∞ performance objectives I: robust performance analysis. *IEEE Trans. Autom. Control*, vol. 39, no. 8, pp. 1564–1574.

[4] Doyle, J., Zhou, K., Glover, K. and Bodenheimer, B. (1994). Mixed H_2 and H_∞ performance objectives II: optimal control. *IEEE Trans. Autom. Control*, vol. 39, no. 8, pp. 1575–1587.

[5] Scherer, C. (1995). Mixed H_2/H_∞ control. In A. Isidori (Ed.), *Trends in Control. A European Perspective*, pp. 173–216. Berlin: Springer-Verlag.

[6] Chen, X. and Zhou, K. (2001). Multiobjective H_2/H_∞ control design. *SIAM J. Control Optim.*, vol. 40, no. 2, pp. 628–660.

[7] Takahashiy, R.H.C., Palharesz, R.M., Dutray, D.A. and Goncalves, L.P.S. (2004). Estimation of Pareto sets in the mixed H_2/H_∞ control problem. *Int. J. Systems Science*, vol. 35, no. 1, pp. 55-67.

[8] Molina-Cristba, A., Griffin, I.A., Fleming, P.J. and Owens, D.H. (2006). Linear matrix inequalities and evolutionary optimization in multiobjective control. *Int. J. Systems Science*, vol. 37, no. 8, pp. 513–522.

[9] Mäkilä, P.M. (1989). On muptiple criteria stationary linear quadratic control. *IEEE Trans. Autom. Control*, vol. AC-34, no.12, pp. 1311–1313.

[10] Khargonekar, P.P. and Rotea, M.A. (1991). Muptiple objective optimal control of linear systems: the quadratic norm case. *IEEE Trans. Autom. Control*, vol. AC-36, no. 1, pp. 14–24.

[11] Hindi, H.A., Hassibi, B. and Boyd, S.P. (1998). Multiobjective H_2/H_∞-optimal control via finite dimensional Q-parametrization and linear matrix

inequalities. In *Proc. 1998 American Control Conf.*, Philadelphia, USA, pp. 3244–3249.

[12] Scherer, C., Gahinet, P. and Chilali, M. (1997). Multiobjective output-feedback control via LMI optimization. *IEEE Trans. Autom. Control*, vol. 42, no. 7, pp. 896–911.

[13] Balandin,D.V., Bolotnik, N.N. and Pilkey, W.D. (2001). *Optimal Protection from Ipmact, Shock, and Vibration*. Amsterdam: Gordon and Breach Science Publishers.

[14] Pilkey, W.D., Balandin, D.V., Bolotnik, N.N., Crandall, J.R. and Purtsezov. S.V. (2010). *Injury Biomechanics and Control: Optimal Protection from Ipmact*. Hoboken, New Jersey: John Wiley and Sons, Inc.

[15] Iwasaki, T. (1996). Robust performance analysis for systems with structured uncertainty. *Int. J. Robust and Nonlin. Control*, vol. 6, pp. 85–99.

[16] Balandin D.V. and Kogan, M.M. (2008). LMI based output-feedback controllers: γ_0-optimal versus linear-quadratic. In *Proc 17th IFAC World Congress, Seoul, Korea*, pp. 9905-9909.

[17] Kogan, M.M. (2016). Optimal discrete-time H_∞/γ_0 filtering and control under unknown covariances. *Int. J. Control*, vol. 89, no. 4, pp. 691–700.

[18] Khargonekar, P.P., Nagpal, K.M. and Poolla, K.R. (1991). H_∞ control with transients. *SIAM J. Control Optim.*, vol. 29, no. 6, pp. 1373–1393.

[19] Nagpal, K.M. and Khargonekar, P.P. (1991). Filtering and smoothing in an H_∞ setting. *IEEE Trans. Autom. Control*, vol. 36, no. 2, pp. 152–166.

[20] Uchida, K., Kojima, A. and Fujita, M. (1997). H_∞ control attenuating initial-state uncertainties. *Int. J. Control*, vol. 65, no. 2, pp. 245–252.

[21] Balandin, D.V. and Kogan, M.M. (2010). LMI based H_∞-optimal control with transients," *Int. J. Control*, vol. 83, no. 8, pp. 1664–1673.

[22] Wilson, D.A. (1989). Convolution and Hankel operator norms for linear systems. *IEEE Trans. Autom. Control*, vol. 34, no. 1, pp. 94–97.

[23] Grimble, M.J. (1990). Relationship between the trace and maximum eigenvalue norms for linear quadratic control design. *IEEE Trans. Autom. Control*, vol. 35, no. 10, pp. 1176–1181.

[24] Zhu, G., Skelton, R. (1991). Mixed L_2 and L_∞ problems by weight selection in quadratic optimal control. *Int. J. Control*, vol. 53, no. 5, pp. 1161–1176.

[25] Rotea, M.A. (1993). The generalized H_2 control problem. *Automatica*, vol. 29, no. 2, pp. 373–385.

[26] Basar, T. and Bernhard, P. (1995). *H_∞-Optimal Control and Related Minimax Design Problems: a Dynamic Game Approach*. Boston: Birkhäuser.

[27] Gahinet, P. and Apkarian, P. (1994). A linear matrix inequality approach to H_∞ control. *Int. J. Robust and Nonlin Control*, vol. 4, pp. 421–448.

[28] Iwasaki, T. and Skelton, R.E. (1994). All controllers for the general H_∞ control problem: LMI existence conditions and state space formulas. *Automatica*, vol. 30, pp. 1307–1317.

In: Robust Control: Systems, Theory and Analysis ISBN: 978-1-53610-826-2
Editor: Sharon Bennett

Chapter 3

On the Geometry of the Attractive Ellipsoids Method: Applications to the Robust Control Design of Switched Systems

Vadim Azhmyakov[*]
Department of Basic Sciences, Universidad de Medellin,
Medellin, Colombia

Abstract

This contribution deals with a robust control design for general switched affine control systems. Dynamical models under consideration are described by ordinary differential equations involving a switching mechanism and in the presence of bounded uncertainties. The design procedure we analyse is based on the newly elaborated attractive ellipsoids method ([32]). The stability and robustness of the resulting closed-loop system involves an abstract Clarke stability theorem and a theoretic extension of the celebrated Lyapunov-type methodology. A short discussion on the obtained analytic results and possible applications and extensions is also included.

Keywords: Switched Dynamic Systems, Robust Control, Attractive Ellipsoids, Invariant Sets

[*]E-mail address: vazhmyakov@udem.edu.co.

1. Introduction

Our paper is devoted to a specific geometric estimation associated with the extended invariant (attractive) ellipsoids techniques. Recall that the conventional method of the invariant ellipsoids (see e.g., [25, 19, 20]) constitutes a powerful theoretical and numerical approach to the problem of control design of various types of dynamic systems with uncertainties (see [17, 23, 28, 31, 32]). Recently, the problem of effective methods for the robust control design of switched systems has attracted a lot of attention, thus both theoretical results and engineering applications were developed, (see e.g. [1, 2, 3, 4, 5, 6, 8, 11, 12, 9, 10, 15, 16, 17, 23, 24, 26, 27, 28, 29, 30, 31, 32, 33]). Useful control design strategies are based on modifications of the classic invariant ellipsoids approach. Let us recall that a set in the state space of a switched dynamic system is said to be positively invariant if any trajectory initiated in this set remains inside the set at all future time instants. One also can also define so called flow-invariant set associated with a dynamic system under consideration. We refer to [14] for the necessary mathematical details. If we replace the above invariancy by an attractive property for the dynamic systems under consideration, we obtain a concept of an attractive ellipsoid (see e.g., [23, 32].

The general question of existence of an invariant set for an arbitrary dynamical system constitutes in general a sophisticated mathematical problem. For the class of systems under consideration this problem is "automatically" solved in the framework of the attractive ellipsoids method. Assuming the existence of an invariant set the next challenging problem is a constructive geometric characterization of this invariant set. In the case of a switched dynamic system it is important to specify an invariant set constructively. The approach we propose makes it possible to obtain an invariant set for the given systems in the form of an ellipsoid (see [17, 25, 19, 20, 28, 29, 30, 31, 32, 33]). An ellipsoidal invariant is evidently a geometrically simple region that can be defined as a "practical stability" region for the given system. A concrete construction of this invariant (or attractive) ellipsoid and the corresponding problem of synthesis of a robust feedback controller involves an auxiliary optimization technique, namely, LMIs (Linear Matrix Inequalities) constrained minimization (see [13, 26, 27, 28, 30]). Note that this LMI-based approach is a direct consequence of the Lyapunov analysis based on quadratic Lyapunoiv-type (storage) functions.

In this paper we propose a new estimation of an ellipsoidal invariant set for the class of switched systems. This estimation is based on the geometrical properties of the dynamic systems under consideration and on an abstract result from [14]. Our analysis involves the basic linearization techniques for the class of switched control systems with nonlinearly affine structure. Our aim is to give an alternative interpretation of the newly obtained results from [17, 24, 28, 32]). A possible additional characterization we propose here extends the initially numeric aspects of the elaborated computational techniques. Let us also note that the conventional invariant/attractive ellipsoid method was developed for stationary quasi-linear systems (see [1, 4, 5, 26, 28, 29, 30, 31, 32]). The stationarity assumption is a significant hypothesis that evidently has a restrictive nature. As a direct consequence of this restrictive assumption one get a suitable reformulation in the form of an LMI-constrained optimization problem. The theoretic estimation of the region of practical stability we propose in this work is free from the stationarity hypothesis and is adapted to the non-stationary affine systems. The geometrical approach to the robust control design of switched systems we propose makes it possible to construct an a priory estimation of the practical stability region of the closed-loop system.

The remainder of our paper is organized as follows. Section 2 contains the problem formulation and detailed description of the dynamic model under consideration. In Section 3 we recall useful concepts and prove some auxiliary mathematical results. Section 4 is devoted to a geometrical characterization of the invariant ellipsoid method for the linearized switched system. We also discuss the resulting LMIs-based control design procedure for the obtained piecewise-linead dynamic model. The robust control synthes is derived here as a formal consequence of the invariancy condition. In Section 5 we propose a robust control design for the initially given nonlinearly affine switched system. This feedback-type control synthesis is based on the robust control design of the associated linear system (considered in Section 4). We use the invariancy property of the ellipsoid designed for the linearized system and propose a constructive estimation and control design for the original switched system. Section 6 summarizes our paper.

2. Problem Formulation and Motivation

Let us introduce the basic dynamic systems concept. In this paper, we restrict our consideration to nonlinearly affine switched systems with the location transitions triggered by time (by time instants).

$$\dot{x}(t) = \sum_{i=1}^{\infty} \chi_{[t_{i-1},t_i)}(t)\, [\, f_{q_i}(x(t)) + B_{q_i}(t)u(t)\,] + \xi(t) \;\text{ a.e. on } \mathbb{R}_+,$$
$$x(0) = x_0, \tag{1}$$

where $x_0 \in \mathbb{R}^n$ is a fixed initial state, all functions $f_{q_i} : \mathcal{R} \to \mathbb{R}^n$, where q_i are from an index set $\mathcal{Q}$, are measurable with respect to variable $t \in \mathbb{R}_+$ and uniformally Lipschitz continuous on an open set $\mathcal{R} \subseteq \mathbb{R}^n$. We also assume that for every $t \in \mathbb{R}_+$ the values $B_q(t) = (b^q_{k,j}(t))^n_m$ are $n \times m$ matrices with continuous components $b^q_{kj}(\cdot)$ for $q \in \mathcal{Q}$. The initial dynamic system contains the uncertainties $\xi(\cdot)$ that are assumed to be uniformly bounded

$$\sup_{t \in \mathbb{R}_+} ||\xi(t)||^2 \leq M.$$

We consider system (1) associated with an infinitely (countable) number of switching times $\{t_i\}$, $i = 1, ..., \infty$, where

$$0 = t_0 < t_1 < ... < t_{i-1} < t_i < ... \; .$$

and denote by $\chi_{[t_{i-1},t_i)}(\cdot)$ the characteristic function of the disjunct time intervals of the type $[t_{i-1}, t_i)$, $i = 1, ..., r$. Motivated by numerous engineering applications, let us next consider system (1) over a control set $\mathcal{U}$ of essentially bounded measurable control inputs. Niote that the last condition implies that the corresponding trajectories have bounded derivatives. We assume that for every admissible control function $u(\cdot) \in \mathcal{U}$ the initial value problem (1) has an absolutely continuous (and continuously prolongable) solution $x^u(\cdot)$. However, due to the presence of a nonlinearity $A_{q_i}(t, x)$ in the above equation, the global existence of a solution to (1) is not guaranteed, unless there are some additional conditions imposed. We refer to [25, 30] for the corresponding details. We next call the above technical assumptions associated with the initial value problem (1) are called the "basic systems assumptions".

Control system (1) can be considered as a suitable formal dynamical model for various classes of switched and hybrid systems. We refer to[6, 7, 22, 34] for some concrete definitions and concepts of hybrid/switched control systems. Note that for some classes of the above systems the possible switching mechanisms are formalized by different types of the characteristic (transition) functions. For example, some classes of hybrid systems with autonomous location transitions (considered in [6, 34]) have the switching times t_i depending on the trajectory $x^u(\cdot)$. The general switched systems with controlled switchings are introduced in [21]. We also refet to [7, 35] for so called switched-mode control systems.

In parallel with (1) we next introduce the corresponding linearized control system. In fact we deal with the locations-based (sequencial) linearization of the basic nonlinear system (1)

$$\begin{aligned} \dot{y}(t) &= \sum_{i=1}^{\infty} \chi_{[t_{i-1},t_i)}(t) \; [\; f_{q_i,x}(x^u(t))y(t) + B_{q_i}(t)v(t) \;] + \xi(t), \\ y(0) &= 0, \end{aligned} \tag{2}$$

where $u(\cdot) \in \mathcal{U}$ and $x^u(\cdot)$ is the absolutely continuous solution to the initial system (1) generated by the admissible $u(\cdot)$. Note that the auxiliary linear system (1) is studied for a fixed admissible control function $u(\cdot)$. This preselected control strategy is usually called the reference control. We also make the following additional technical assumptions related to (2), namely, we suppose that every pair $(A_{q_i}(t), B_{q_i}(t))$, where

$$A_{q_i}(t) := f_{q_i,x}(x^u(t)),$$

is controllable for every $t \in \mathbb{R}_+$ and every $q_i \in \mathcal{Q}$. Evidently, system (2) is a piecewise-linear system. Let us now restrict the given class $\mathcal{L}$ of admissible control functions for the linearized system (2). In this paper we deal with a family of locally Lipschitz feedback type control functions $w(\cdot,\cdot)$ such that

$$w(t, y^v(t)) = v(t)$$

(see (2), where $y^v(\cdot)$ is a solution to (2) generated by an admissible (open-loop) control $v(\cdot)$. For each $u(\cdot) \in \mathcal{U}$ and $w(\cdot,\cdot) \in \mathcal{L}$ the initial value problem (2) has a unique solution denoted by $y^w(\cdot)$. We next restrict our consideration to a subclass of the introduced admissible feedback controls and consider the concrete control design of the following type

$$\begin{aligned} w(t, y(t)) &= \sum_{i=1}^{\infty} \chi_{[t_{i-1},t_i)}(t) K_{q_i}(t) y(t), \\ K_{q_i}(t) &\in \mathbb{R}^{m \times n}, \; t \in \mathbb{R}_+ \, , \; q_i \in \mathcal{Q}, \end{aligned}$$

where $K_{q_i}(\cdot)$ are gain matrix-functions associated with the given locations $q_i \in \mathcal{Q}$. These unknown gain matrices constitutes in fact a feedback-type control design (from the class of piecewise-linear functions) for the given switched system. The closed-loop form of (2) can now be written as follows:

$$\begin{aligned} \dot{y}(t) &= \sum_{i=1}^{\infty} \chi_{[t_{i-1},t_i)}(t)\, [\, f_{q_i,x}(x^u(t)) + B_{q_i}(t) K_{q_i}(t))\,] y(t) + \xi(t), \\ y(0) &= 0, \end{aligned} \tag{3}$$

From the qualitative point of view the chosen systems design can be described as follows: the generated trajectories of (3) needs to stay in a "small" ellipsoidal region with a center at the origin

$$\mathcal{E} := \{ y \in \mathbb{R}^n \mid y^T P y \leq 1 \},$$

where P is a positive defined symmetrical $n \times n$ matrix. Let us also refer to [31] for the related definitions and basic concepts.

The aim of our contribution is to give a geometric baesed estimation for the known invariant/attractive ellipsoid approach. Note that various types of linearized dynamic models have been long time recognized as a powerful tool for solving stability and stabilization problems (see e.g., [37]. The above-mentioned geometric stability criterion for the linearized system (2) can routinely be used in the design procedure of a robust feedback controller for the original nonlinear system (1). Let us also note that the resulting uncertain piecewise-linear system (3) is a non-stationary system. Recall that the non-stationarity hypothesis constitutes a crucial restriction for the direct application of the conventional (LMI-based) attractive ellipsoids techniques [4, 17, 23, 24, 28, 29, 32].

3. A Characterization of the Linearized System

We now formulate an analytic result that establishes the "quality" of the linear-type approximations (2) of the original switched system (1). We use here the standard notation $\mathbb{L}^{\infty}_s$, where $s \in \mathbb{N}$, for a standard Lebesgue space of measurable and essentially bounded s-dimensional vector functions defined on $\mathbb{R}_+$. Moreover, $x^{u+v}(\cdot)$ denotes the trajectory of the original system (1) generated by the admissible control $u(\cdot) + v(\cdot)$.

Theorem 1. *Assume that the initial system* (1) *satisfies all the basic technical assumptions. Then there exists a function* $o : \mathbb{R}_+ \to \mathbb{R}_+$ *such that*

$$s^{-1}o(s) \to 0 \text{ as } s \downarrow 0$$

and the following estimation holds:

$$||x^{u+v}(\cdot) - (x^u(\cdot) + y^v(\cdot))||_{\mathbb{L}_n^\infty} \leq o(||v(\cdot)||_{\mathbb{L}_m^\infty})$$

for all $u(\cdot) \in \mathcal{U}$ *and* $v(\cdot) \in \mathbb{L}_m^\infty$.

Proof. Let $u(\cdot) \in \mathbb{L}_m^\infty$. For a function $w(\cdot,\cdot) \in \mathcal{L}$ we evidently get $v(\cdot) \in \mathbb{L}_m^\infty$. Let $I \subset \mathbb{R}_+$ be a time interval. The basic comparison result with the comparison functions

$$\begin{aligned} z(t) &:= x^u(t) + y^v(t), \\ \psi(t,x) &= \sum_{i=1}^{\infty} \chi_{[t_{i-1},t_i)}(t)\,[f_{q_i}(x) + B_{q_i}(t)u\,] + v(t), \end{aligned}$$

for $t \in I$ implies

$$\begin{aligned} &||x^{u+v}(\cdot) - (x^u(\cdot) + y^v(\cdot))||_{\mathbb{L}_n^\infty} \leq e^C \int_I ||\dot{x}^u(t) + \dot{y}^v(t) - \\ &\sum_{i=1}^{\infty} \chi_{[t_{i-1},t_i)}(t)\,[f_{q_i}(x^u(t) + y^v(t)) - B_{q_i}(t)(u(t) + v(t))\,]||dt \leq \\ &e^C \int_I \sum_{i=1}^{\infty} \chi_{[t_{i-1},t_i)}(t)\,[\,||\langle (f_{q_i,x}(x^u(t)), B_{q_i}(t)), (y^v(t), v(t))\rangle - \\ &(f_{q_i}(x^u(t) + y^v(t)) + B_{q_i}(t)v(t) - f_{q_i}(x^u(t))||\,]dt. \end{aligned} \tag{4}$$

Here C is a constant.

Using the component-wise variant of the Mean Value Theorem, we next deduce

$$\begin{aligned} &\sum_{i=1}^{\infty} \chi_{[t_{i-1},t_i)}(t)\,[f_{q_i}^i(x^u(t) + y^v(t)) + B_{q_i}(t)(u(t) + v(t)) - (f_{q_i}^i(x^u(t)) + B_{q_i}(t)u(t))] = \\ &\sum_{i=1}^{\infty} \chi_{[t_{i-1},t_i)}(t)\langle (f_i)_{q_i,x}(x^u(t) + g(t)), B_{q_i}(t))(y^v(t), v(t))\rangle \end{aligned}$$

for a suitable bounded $g(\cdot)$. Here $i = 1, ..., n$, $j = 1, ..., m$. The Lipschitz continuity of functions $f_q(\cdot)$, $q \in \mathcal{Q}$ on a bounded set $\mathcal{R}$ implies the existence of a function $o_1 : \mathbb{R}_+ \to \mathbb{R}_+$ such that $s^{-1}o_1(s) \to 0$ as $s \downarrow 0$ and

$$\sum_{i=1}^{\infty} \chi_{[t_{i-1},t_i)}(t) \, [\, ||\langle (f_x(x^u(t)), B(t)), (y^v(t), v(t)) - \\ (f(x^u(t) + y^v(t)) + B(t)v(t) - f(x^u(t))) \,] \leq o_1(||v(\cdot)||_{\mathbb{L}^\infty_m}) \tag{5}$$

for all $t \in I$. Using (4) and (5), we finally deduce the final inequality

$$||x^{u+v}(\cdot) - (x^u(\cdot) + y^v(\cdot))||_{\mathbb{L}^\infty_n} \leq o(||v(\cdot)||_{\mathbb{L}^\infty_m}),$$

where $o(s) := e^C o_1(s)$. Since the last inequality is true for every $I \subset \mathbb{R}_+$, it is also true for all $t \in \mathbb{R}_+$. The proof is finished. □

We next will apply Theorem 1 and deduce an ellipsoidal estimation associated with the robust control design procedure for the original nonlinear system (1).

Following the basic concept introduced in [14], we call a smooth manifold S in an Euclidean space a flow-invariant for the dynamic system

$$\begin{aligned} \dot{z}(t) &= \phi(z(t)), \ t \in \mathbb{R}_+, \\ z(0) &= 0 \end{aligned} \tag{6}$$

if $z(t) \in S$ for all $t \geq T \in \mathbb{R}_+$. The next abstract theorem gives a general criterion of the property to be flow-invariant.

Theorem 2. *A smooth manifold S is flow-invariant for system* (6) *iff $\phi(x)$ belongs to the tangent space T_S of S for all x from the given* Eucledian *space.*

Note that Theorem 2 has a very natural geometrical interpretation. The formal proof of this abstract result is based on an extended Lyapunov-type technique (see [14] for details).

Recall that a set $\mathcal{D}$ in the state space of system of a dynamic system is said to be (positively) invariant if an admissible trajectory initiated in this set remains inside the set at all future time instants. Let us denote by $\Omega(z(0))$ the (positive) limit set of system (3) (the set of all positive limit points, see e.g., [16, 25, 18]). We now give a related Lyapunov set stability concept (see [18, 25])

Definition 1. *A compact invariant set* $\mathcal{D} \subset \mathbb{R}^n$ *of the closed-loop dynamic system* (3) *that possesses a finite is called asymptotically* Lyapunov *stable if,* $\Omega(z(0)) \subset \mathcal{D}$ *and*

- *for all* $\epsilon > 0$ *there exists* $\delta_1 > 0$ *such that* $\text{dist}[z(0), \mathcal{D}] \leq \delta_1$ *implies* $\text{dist}[z(t), \mathcal{D}] \leq \epsilon$ *for all* $t \in \mathbb{R}_+$ *(the* Lyapunov *stability of the set);*
- *there exists* $\delta_2 > 0$ *such that* $\text{dist}[z(0), \mathcal{D}] \leq \delta_1$ *implies* $\lim_{t\to\infty} \text{dist}[z(t), \mathcal{D}]$ *(the attraction property of the set).*

Here

$$\text{dist}[z; \mathcal{D}] := \min_{\tilde{z}\in\mathcal{D}} ||z - \tilde{z}||_{R^n}$$

is the usual Euclidean distance between a point $z \in \mathbb{R}^n$ and $\mathcal{D}$. As mentioned above, the existence and an expected constructive characterization of an invariant set for a general dynamical system of the type (6) are sophisticated analytical questions. Our aim is to specify an invariant set for the closed-loop piecewise-linear system (3) in the form of an ellipsoid $\mathcal{E}$ introduced in Section II. Applying the classic definition of an asymptotically stable set, namely, Definition 1 we now introduce the main concept of the attractive ellipsoid.

Definition 2. *We call* $\mathcal{E}$ *an attractive ellipsoid for system* (3) *if it is an asymptotically stable invariant set of* (3).

It is easy to see that an attractive ellipsoid (if exists) for the switched piecewise-linear system (3) is determined by the matrix P and also by the family of designed gain matrices K_q, where $q \in \mathcal{Q}$. The chosen gain matrices K_q determine in fact the resulting dynamic behavior of the complete state vector y(t) such that the basic inequality

$$yT(t)Py(t) \leq 1, \; t \in R_+$$

holds. The last condition can be considered in an exact or in any approximative sense. This restricted dynamic behavior of the piecewise-linear system (3) closed by the linear-type feedback can be interpreted as a stability of the resulting system in the practical sense.

4. A Geometrical Attractive Ellipsoids Method for the Linearized Switched System

In this section we use the previously discussed facts and give a geometric interpretation of the attactive ellipsoids method for the piecewise-linear switched system (3). Let us introduce an additional dynamic variable $\theta(\cdot)$, $\theta(0) = 0$ and determine the smooth manifold in the extended Euclidian state space $\mathbb{R}^{n+1}$

$$S := \{z \in \mathbb{R}^{n+1} \mid y^T(t)Py(t) - 1 + \theta(t) = 0\}.$$

Here $z := (y, \theta)^T$ and $y(\cdot)$ is a solution to (3). Introduce the following additional notation

$$h := z^T P z - 1 + \theta.$$

The necessary and sufficient condition for invariance of the manifold S determined above can be now formlized using Theorem 2:

$$\begin{aligned} &\langle \nabla h(z), \phi(z) \rangle = 0, \\ &\theta(t) \geq 0. \end{aligned} \tag{7}$$

By ∇h we denote the gradient of the function h introduced above. The vector field ϕ corresponds to the right hand side of the following system of equations

$$\begin{aligned} &\dot{y}(t) = \sum_{i=1}^{\infty} \chi_{[t_{i-1},t_i)}(t) \left[\, (f_{q_i,x}(x^u(t)) + B_{q_i}(t)K_{q_i}(t))y(t) \, \right] + \xi(t), \\ &\dot{\theta}(t) = -2y^T(t)P\dot{y}(t), \\ &y(0) = 0, \;\; \theta(0) = 0. \end{aligned} \tag{8}$$

We now use the dynamics of system (3) and rewrite the second differential equation from (8) in the equivalent form:

$$\dot{\theta}(t) = -2y^T(t)P\Big[\sum_{i=1}^{\infty} \chi_{[t_{i-1},t_i)}(t) \left[\, (f_{q_i,x}(x^u(t)) + B_{q_i}(t)K_{q_i}(t))y(t) \, \right] + \xi(t)\Big]. \tag{9}$$

The above observations can be summarized as the following result.

Theorem 3. *The ellipsoidal set $\mathcal{E}$ is an invariant set for the closed-loop system* (3) *if and only if the variable $\theta(t)$ determined by the initial-value problem* (8) *is a non-negative for all $t \in \mathbb{R}_+$.*

Proof. Compute the necessary scalar product in (7) $\langle \nabla h(z), \phi(z) \rangle \big|_{z=z(t)}$. Evidently, we have

$$\langle (2y^T(t)P, 1), (\dot{y}(t), -2y^T(t)P\dot{y}(t)) \rangle = 2y^T(t)P\dot{y}(t) - 2y^T(t)P\dot{y}(t)) = 0.$$

Evidently, the first condition from (7) is true. Therefore, (7) is reduced to the second condition, namely, to the non-negativity of the variable $\theta(t)$ determined by system (8). The proof is completed. □

The formal solution of the initial-value problem on a time interval I for the artificial variable $\theta(\cdot)$ can be expressed as follows:

$$\theta(t) = -\int_I [2y^T(t)P \sum_{i=1}^{\infty} \chi_{[t_{i-1},t_i)}(t)\ [\ f_{q_i,x}(x^u(t)) + B_{q_i}(t)K_{q_i}(t)\ ]y(t) + y^T(t)P\xi(t)]\, dt.$$

Consequently, the non-negativity condition for $\theta(t)$ established in Theorem 2 implies the inequality

$$\int_I [y^T(t)P \sum_{i=1}^{\infty} \chi_{[t_{i-1},t_i)}(t)\ [\ f_{q_i,x}(x^u(t)) + B_{q_i}(t)K_{q_i}(t))\ ]y(t) + y^T(t)P\xi(t)]\, dt \leq 0. \quad (10)$$

Taking into acount the robust control applications we are naturally interested to construct an attractive ellipsoid $\mathcal{E}$ of a minimal size. This requirement and the corresponding selection of the ellipsoid matrix P need to be adequately formalized. Let us recall the usual formalization procedure and the associated parameter selection in the form of a specific LMI-constrained minimization problem (see e.g., [13]). For the case of stationary dynamic systems one can determine a minimal-size matrix P and design the corresponding gain matrices K_{q_i} using the following optimization problem (see [4, 5, 32, 33]):

$$\begin{aligned} &\text{minimize } \operatorname{tr}(P^{-1}) \\ &\text{subject to } P^T = P,\ P > 0, \\ &\{K_{q_i}(t)\} \in \mathcal{K} \text{ for } t \in \mathbb{R}_+, \end{aligned} \quad (11)$$

where $\mathcal{K} \subset \mathbb{R}^{m \times n}$ is the set of admissible matrices that ensure invariance of attractive ellipsoid $\mathcal{E}$ for the given switched dynamic system (3). Evidently, a constructive description of the set $\mathcal{K}$ constitutes a sophisticated formal problem. This problem can be constructively solved in the case of stationary dynamic systems and moreover, under some additional restrictive conditions. The same is also true for the originally given (non-linearized) system. The switched control systems discussed above, namely, dynamic models (1) and (3) have a strongly non-stationary nature. This fact is an immediate consequence of the presence of a time-driven switching mechanism. Therefore, in the case of a switched dynamics given by the nonlinearly affine system (1) and by the piecewise-linear system (3) the non-stationary nature of the resulting systems make it impossible a direct application of the LMI-based approach. Roughly speaking one can not give a formal description of the set $\mathcal{K}$ of admissible gain matrices in the form of a unique LMI. The restrictions set $\mathcal{K}$ in (11) evidently has a dynamic structure and contains time-depending gain matrices. On the other side a solution of the time-depending optimization problem (11) involves a very massive calculation related to every time instant.

The qualitative analysis and the explanations given above imply the necessity to develope a conceptually new approach to the parameter selection of an attractive ellipsoid (matrix P). Then same is also true with respect to the resulting feedback control design (matrices $K_{q-i}(t)$). We use here a guaranteed approach and select (a priory) a suitable matrix $\hat{P}$ that involves a relative "small" invariant set for the closed-loop system (3).

Theorem 4. *For a given admissible ellipsoid matrix $\hat{P}$ with $\hat{P} = \hat{P}^T > 0$ the suitable gain martices $\{K_{q_i}(t)\} \in \mathcal{K}$, $q_i \in \mathcal{Q}$, $t \in \mathbb{R}_+$ can be found as solutions of the following* LMI

$$\hat{P}\sum_{i=1}^{\infty} \chi_{[t_{i-1},t_i)}(t)\ [\ f_{q_i,x}(x^u(t)) + B_{q_i}(t)K_{q_i}(t)\ ] + \sqrt{M}||\hat{P}|| \times E \leq 0, \quad (12)$$

where E is a $n \times n$ unit matrix.

Proof. The first summand in (12) evidently coincides with the matrix of the first summand in the integrand in (10). Let us now estimate the expression $y^T(t)P\xi(t)$ from (10). We have

$$(y^T(t)P\xi(t))^2 = y^T(t)P\xi(t)\xi^T(t)Py \leq ||\hat{P}||^2 y^T(t)\xi(t)\xi^T(t)y(t)$$

For the Frobenius norm $||\xi(t)\xi^T(t)||$ of the matrix $\xi(t)\xi^T(t)$ we obtain (see Section 2)

$$||\xi(t)\xi^T(t)|| = ||\xi(t)||^2 \leq \sup_{t\in\mathbb{R}_+} ||\xi(t)||^2 \leq M.$$

Therefore,

$$(y^T(t)P\xi(t))^2 \leq y^T(t)(M||\hat{P}||^2 E)y(t)$$

and the matrix condition (12) implies the integral inequality (10). Using (10) and Theorem 2, we deduce that the obtained LMI (12) determines the gain matrices $\{K_{q_i}(t)\} \in \mathcal{K}$. That means a set of solutions $\{K_{q_i}(t)\}$ of (12) for $t \in \mathbb{R}_+$ involves the invariance of the given ellipsoid $\mathcal{E}$. The proof is completed. □

The LMI-type matrix inequalities (12) provide a theoretic fundament for an implementable computational procedure. Note that Theorem 3 express only a sufficient condition for the inclusion $\{K_{q_i}(t)\} \in \mathcal{K}$, $q_i \in \mathcal{Q}$, $t \in \mathbb{R}_+$ (for the invariance of $\mathcal{E}$). Solutions to (12) also depend on the prescribed reference state $x^u(\cdot)$ associated with the initially given switched system (1).

5. A Geometrical Attractive Ellipsoids Method for the Originally Given Switched System

We now back to the original nonlinearly affine control system (1) and propose an admissible control design of the following srtucture:

$$u(t) := u^{ref}(t) + \sum_{i=1}^{\infty} \chi_{[t_{i-1},t_i)}(t) w_{q_i}(t, y(t)),\ w_{q_i}(t, y(t)) := K_{q_i}(t)y(t), \tag{13}$$

where $y(\cdot)$ is a solution to the linearized system (3) closed by the piecewise-linear feedback $w(t, y) = \sum_{i=1}^{\infty} \chi_{[t_{i-1},t_i)}(t)K_{q_i}(t)y$ and $u^{ref}(\cdot)$ is a selected tracking (reference) control associated with the given system (1). We assume that $u^{ref}(\cdot)$ is an essentially bounded measurable function (an admissible control from $\mathcal{U}$, see Section 2). This assumption evidently guarantee $u(\cdot) \in \mathcal{U}$. An absolutely continuous solution of the initial system (1) generated by $u^{ref}(\cdot)$ is denoted by x^{ref}. Note that $y(\cdot)$ in the proposed feedback-type control design is a solution of the linearized system (3) generated by the piecewise-linear feedback control $w(t, y)$ (see Section 4). The chosen combined control function $u(\cdot)$ proposed in (13) needs to possesses the robustness property in the following sense: the closed-loop variant of system (1)

$$\begin{aligned} \dot{x}(t) &= \sum_{i=1}^{\infty} \chi_{[t_{i-1},t_i)}(t) \left[f_{q_i}(x(t)) + B_{q_i}(t)u^{ref}(t) + B_{q_i}(t)K_{q_i}(t)y(t) \right] + \xi(t) \\ x(0) &= x_0, \end{aligned} \tag{14}$$

admits an invariant (attractive) ellipsoid $\mathcal{E}_0$, $0 \in \mathcal{E}_0$. We also use here the notation of practical stability as an alternative terminology. Note that the feedback-type control design (13) depends on the state vector $y(\cdot)$ of the correspondingly linearized system (3) replacing $u(t) = u^{ref}(t),\ t \in \mathbb{R}_+$. Our aim is to establish the ellipsoidal-type stability (practical stability) of the original nonlinear system (1) using the piecewise-linear robust control design developed for (3). For a given admissible matrix $\hat{P}$ and taking into consideration the designed attractive ellipsoid $\mathcal{E}$ of the piecewise-linear switched system (3) we next give a constructive estimation of the attractive ellipsoid $\mathcal{E}_0$ associated with the nonlinearly affine switched system (1). We use the abstract results from Section 3, namely, Theorem 1 for this purpose.

Theorem 5. *Consider an initially given switched system (1) that satisfies all the technical assumptions from Section 2 and the corresponding variant (4) of this system closed by the feedback-type control (13). Let $\hat{P}$ be a selected admissible matrix for $\mathcal{E}$ and the gain matrices $\{K_{q_i}\}$ are determined by (12). Assume that the reference trajetory $x^{ref}(\cdot)$ generated by $u^{ref}(\cdot) \in \mathcal{U}$ is uniformly bounded $||x^{ref}(t)|| \leq \chi,\ t \in \mathbb{R}_+$. Then the attractive ellipsoid $\mathcal{E}_0$ associated with the closed-loop variant (15) of the nonlinearly affine switched system (1) admits the following estimation:*

$$\mathcal{E}_0 := (\max_{q_i \in Q}\{||K_{q_i}(\cdot)||\} + 1)\ \mathcal{E} + \chi, \tag{15}$$

Proof. Applying Theorem 1, we first obtain the following estimation

$$||x^u(\cdot) - (x^{ref}(\cdot) + y(\cdot))||_{\mathbb{L}_n^\infty} \leq o(\max_{q_i \in Q}\{||K_{q_i}(\cdot)y(\cdot)||_{\mathbb{L}_m^\infty}\}).$$

The above inequality implies the next inequalities

$$\begin{aligned} &||x^u(\cdot) - x^{ref}(\cdot)||_{\mathbb{L}_m^\infty} \leq o(\max_{q_i \in Q}\{||K_{q_i}(\cdot)y^v(\cdot)||_{\mathbb{L}_m^\infty}\}) + ||y(\cdot)||_{\mathbb{L}_n^\infty} \\ &\leq (\max_{q_i \in Q}\{||K(\cdot)||\} + 1)||y(\cdot)||_{\mathbb{L}_n^\infty} \end{aligned}$$

Moreover, we have

$$\begin{aligned} &||x^u(\cdot)|| \leq (\max_{q_i \in Q}\{||K_{q_i}(\cdot)||\} + 1)||y(\cdot)||_{\mathbb{L}_n^\infty} + ||x^{ref}(\cdot)||_{\mathbb{L}_m^\infty} \\ &\leq ||x^u(\cdot)|| \leq (\max_{q_i \in Q}\{||K_{q_i}(\cdot)||\} + 1)||y(\cdot)||_{\mathbb{L}_n^\infty} + \chi \end{aligned} \tag{16}$$

for $t \in \mathbb{R}_+$.

Since $x^{ref}(\cdot)$ is assumed to be bounded and $\mathcal{E}$ is an invariant ellipsoid for (3), we deduce the expected estimation (15). The obtained inequality (16) also implies the invariance property of the ellipsoidal set $\mathcal{E}_0$ associated with the closed-loop system (15). The proof is completed. □

Evidently, the basic Theorem 5 makes it possible to describe constructively an invariant ellipsoidal set $\mathcal{E}_0$ associated with the the sophisticated nonlinearly affine switched system (1) using an invariant ellipsoid $\mathcal{E}$ designed for the linearized system (3). Note that the robust control design $w(t, y)$ for the linearized system (3) has a simple feedback-type structure. Theorem 5 makes it possible to use this relatively easy linear-system based control design as a practically stabilizing control strategy for the sophisticated system (1). Evidently, the size of the ellipsoid $\mathcal{E}_0$ can be "controlled" by the size of $\mathcal{E} \subset \mathcal{E}_0$ and in fact by an adequate choice of the matrix $\hat{P}$.

Finally note that the practical implementation of the proposed control design (13) assumes relatively frequent updates of the following type:

$$u^{ref}(t) \equiv u(t) \text{ for } t \in I_{k+1} \subset \mathbb{R}_+ \setminus I_k,\ k = 0, 1, \dots,$$

where I_k is a "previously given" time interval. This updates are consistent with the general linearization concept for a dynamic system. Recall that a linearization procedure with respect to a reference pair $(u^{ref}(\cdot), x^{ref}(\cdot))$ is acceptable only on a finite time interval. One has to innovate the corresponding references after this time interval is completed.

Conclusion

The presented paper extends the newly elaborated attractive ellipsoids based theoretical approach to a robust control design for a class of nonlinear switched systems. We have considered a general class of non-stationary dynamic models with an affine structure. We firstly discuss a constructive approach to the piecewise-linear (linearized) switched system and design an associated ellipsoidal invariant set. We use here a geometric approach proposed in [2, 32]. We next apply the feedback-type control strategy obtained for the linearized system in the control design procedure related to the original switched system. Our main result contains a formal extension a geometrical interpretation of the attractive ellipsoid method. The general approach we follow makes it possible to obtain a constructive estimation of the ellipsoidal invariant set for the sophisticated non-linear switched system using the relatively easy robust control strategy proposed for linearized switched system.

The presented version of the attractive ellipsoids method in combination with the specific linearization technique provides a basis for possible computational methods in robust control of sophisticated dynamic systems with switching mechanisms. The geometrical approach we propose can also be applied to more general nonlinear control systems, namely, to some classes of hybrid dynamic models. Let us finally note that one can incorporate into the developed analytical framework some alternative types of system uncertainties.

References

[1] J. Alazki and A. Poznyak, Robust output stabilization for a class of nonlinear uncertain stochastic systems under multiplicative and additive noises: The attractive ellipsoid method, *Journal of Industrial and Management Optimization*, vol. 12, 2016, pp. 169 – 186.

[2] V. Azhmyakov, On the geometric aspects of the invariant ellipsoid method: Application to the robust control design, in: *Proceedings of the 50th IEEE Conference on Decision and Control and European Control Conference*, Orlando, USA, 2011, pp. 1353 – 1358.

[3] V. Azhmyakov, A. Poznyak and O. Gonzalez, On the robust control design for a class of nonlinearly affine control systems: The attractive ellipsoid approach, *Journal of Industrial and Management Optimization*, vol. 9, 2013, pp. 579 – 593.

[4] V. Azhmyakov, A. Poznyak and R. Juarez, On the practical stability of control processes governed by implicit differential equations: The invariant ellipsoid based approach, *Journal of The Franklin Institute*, vol. 350, 2013, pp. 2229 – 2243.

[5] V. Azhmyakov, M. Basin and C. Reincke-Collon, Optimal LQ-type switched control design for a class of linear systems with piecewise constant inputs, in: *Proceedings of the 19th IFAC World Congress*, Cape Town, South Africa, 2014, pp. 6976 – 6981.

[6] V. Azhmyakov, J. Cabrera Martinez, A. Poznyak, and R. R. Serrezuela, Optimization of a class of nonlinear switched systems with fixed-levels control inputs, in: *Proceedings of the 2015 American Control Conference*, Chicago, USA, pp. 1770 – 1775 .

[7] V. Azhmyakov and R. Juarez, On the projected gradient methods for switched-mode systems optimization, in: *IFAC-PapersOnLine*, vol. 48, 2015, pp. 181 – 186.

[8] A.E. Barabanov and O.N. Granichin, Optimal controller for linear plants with bounded noise, *Automation and Remote Control*, vol. 45 , no. 5, 1984, pp. 39 – 46.

[9] M. Basin, J. Rodriguez-Gonzalez and L. Fridman, Optimal and robust control for linear state-delay systems, *Journal of the Franklin Institute*, vol. 344, no. 6, 2007, pp. 830 – 845.

[10] M. Basin, P. Rodriguez-Ramirez, S. Ding and S. Dominic, A nonhomogeneous super-twisting algorithm for systems of relative degree more than one, *Journal of The Franklin Institute*, vol. 352, 2015, pp. 1364 – 1377.

[11] M. Bonilla, M. Malabre and V. Azhmyakov, Decoupling of internal variable structure for a class of switched systems, in: *Proceedings of the 2015 European Control Conference*, Linz, Austria, 2015, pp. 1890 – 1895.

[12] M. Bonilla, M. Malabre and V. Azhmyakov, An implicit systems characterization of a class of impulsive linear switched control processes. Part 1: Modeling, *Nonlinear Analysis: Hybrid Systems*, vol. 15, 2015, pp. 157 – 170.

[13] S. Boyd, E. Ghaoui, E. Feron and V. Balakrishnan, *Linear Matrix Inequalities In System And Control Theory*, Philadelphia: SIAM, 1994.

[14] F. Clarke, *Optimization and Nonsmooth Analysis*, Philadelphia, SIAM, 1990.

[15] M.A. Dahleh, J.B. Pearson and J. Boyd, Optimal rejection of persistent disturbances, robust stability, and mixed sensitivity minimization, *IEEE Transactions on Automatic Control*, vol. 33, no. 8, 1988, pp. 722 – 731.

[16] G.J. Duncan and F.C. Schweppe, Control of linear dynamic systems with set constrained disturbances, *IEEE Transactions on Automatic Control*, AC-16, 1971, pp. 411 – 423.

[17] O. Gonzalez, A. Poznyak and V. Azhmyakov, On the Robust control design for a class of nonlinear affine control systems: the invariant ellipsoid approach, in: *Proceedings of the 6th International Conference on Electrical Engineering, Computing Science and Automatic Control*, Toluca, Mexico, 2009, pp. 1 – 6.

[18] H. K. Khalil, Nonlinear Systems, Prentice Hall, Upper Saddle River, NJ, USA, 2002.

[19] A.B. Kurzhanski and P. Varaiya, Ellipsoidal techniques for reachability under state constraints, *SIAM Journal on Control and Optimization*, vol. 45, no. 4, 2006, pp. 1369 – 1394.

[20] A.B. Kurzhanski, V.M. Veliov , *Modeling Techniques and Uncertain Systems*, Birkhäuser, New York, 1994.

[21] D. Liberzon, Switching in Systems and Control, Birkhäuser, Boston, USA, 2003.

[22] J. Lygeros, Lecture Notes on Hyrid Systems, Cambridge University Press, Cambridge, UK, 2003.

[23] M. Mera, F. Castanos and A. Poznyak, Quantised and sampled output feedback for nonlinear systems, *International Journal of Control*, vol. 87, 2014, pp. 2475 – 2487.

[24] M. Mera, A. Polyakov, W. Perruquetti and G. Zheng, Finite-time attractive ellipsoid method using implicit Lyapunov functions, in: Proceedings of the 54th Conference on Decision and Control, Osaka, Japan, 2015, pp. 6892 – 6896.

[25] A. N. Michel, L. Hou and D. Liu, *Stability of Dynamical Systems*, Birkhäuser, New York, 2007.

[26] B.T. Polyak, S.A. Nazin, C. Durieu and E. Walter, Ellipsoidal parameter or state estimation under model uncertainty, *Automatica*, vol. 40, no. 7, 2004, pp. 1171 – 1179.

[27] Polyak, Topunov, Suppression of bounded exogeneous disturbances: output control, *Automation and Remote Control*, vol. 69, no. 5, 2008, pp. 801 – 818.

[28] A. Polyakov and A. Poznyak, Lyapunov function design for finite-time convergence analysis: twisting controller for second-order sliding mode realization, *Automatica*, no. 45, 2009, pp. 444 – 448.

[29] A. Polyakov, Minimization of disturbances effects in time delay predictor-based sliding mode control systems, *Journl of The Franklin institute*, vol. 349, 2012, pp. 1380 – 1396.

[30] A. Poznyak, *Advanced Mathematical Tools for Automatic Control Engineers: Deterministic Techniques*, Elsevier, Amsterdam, 2008.

[31] A. Poznyak, V. Azhmyakov and M. Mera, Practical output feedback stabilization for a class of continuous-time dynamic systems under sample-dada outputs, *International Journal of Control*, vol. 84, 2011, pp. 1408 – 1416.

[32] A. Poznyak, A. Polyakov, and V. Azhmyakov, Attractive Ellipsoids in Robust Control, Birkhäuser, Basel, Switzerland, 2014.

[33] T. Poznyak, I. Chairez and A. Poznyak, Switching robust control for ozone generators using the attractive ellipsoid method, *ISA Transactions*, vol. 53, 2014, pp. 1796 – 1806.

[34] M.S. Shaikh and P.E. Caines, On the hybrid optimal control problem: theory and algorithms, *IEEE Transactions on Automatic Control*, vol. 52, 2007, pp. 1587 – 1603.

[35] Y. Wardi, M.Egerstedt and P.Twu, A controlled precision algorithm for mode-switching optimization, in: *Proceedings of the 51st Conference on Decision and Control*, Maui, USA, 2012, pp. 713 – 718.

[36] J. Zabczyk, *Mathematical Control Theory: an Introduction*, Birkhäuser, Boston, 1995.

[37] V.I. Zubov, *Mathematical Methods for the Study of Automatic Control Systems*, Pergamon Press, New York, 1962.

In: Robust Control: Systems, Theory and Analysis ISBN: 978-1-53610-826-2
Editor: Sharon Bennett

Chapter 4

A Robust Approximated Derivative Action of a PID Regulator to Be Applied in a Permanent Magnet Synchronous Motor Control

Paolo Mercorelli*
Institute of Product and Process Innovation
Leuphana University of Lueneburg, Lueneburg, Germany

Abstract

This paper presented a PID control scheme for a synchronous motor with permanent magnets. PID controllers are the most used controllers in industrial applications and it always presents difficulty to realise the differential part. Moreover, in the last decades a visible effort has been made in the field of sensorless control in which observer and approximation of state variables are used to avoid sensors in the control loop. A polynomial Kalman filter is used to approximate its derivative actions to avoid disturbances and spikes due to the differentiation. The Kalman filter estimates two coefficients at each sampling time, which are necessary to calculate the polynomial of the first order with which the tracking error is intended to be modelled. Simulation results are shown to demonstrate the effectiveness of the approximation.

Keywords: Robust derivative, Synchronous motor, PWM control

*E-mail address: mercorelli@uni.leuphana.de

1. Introduction

Neodymium iron boron (NdFeB) is commercially available for synchronous motors with sinusoidal EMF affordable permanent magnets (PMSM). They are used in some special applications such as robotics, aeronautics and electrical marine propulsion [1], [2]. Synchronous motors with permanent magnets (PMSM) AC traction motors are commonly used in electric or hybrid road vehicles. For railway vehicles, the PMSM as drive motors have not been widely used yet. PMSM were presented at InnoTrans in Berlin 2008 Alstom AGV high-speed train and tram transport ForCity Skoda 15T low floor. The greatest advantage of them is a low volume of PMSM unlike in other types of engines and therefore it allows a direct drive of the wheels. For achieving a desired system performance, advanced control systems are usually required to provide fast and accurate response, quick disturbance recovery and parameter variations insensitivity [3]. One of the most used control strategies is represented by the PID controller. PID controllers have been widely used in industrial applications in the last years because of their simple structure. In particular, derivatives and structures using derivatives, as for instance PD-controllers, are very often used in industrial applications. Such kind of controllers are much used in practical applications of any nature. They are used also in electromagnetic valve control,[4]. Nevertheless, it is known that the derivative part of a PID controller can generate undesired spikes because of the presence of noise. These spikes generate unrobustness in the control loop and in this sense it is important to develop techniques to avoid spikes. In the literature different solutions are proposed to overcome this aspect as in [5] and [6]. In this paper the derivative part of the PID controller is obtained by using a Kalman filter which approximates the error with polynomial approximations. The Kalman filter which is used as an observer is very often applied in control. In the recent decades it has had a great difficulty in controlling a summer sensor generated to achieve a minimum number of the used sensors. Control strategy for permanent magnet drives, in which electromagnetic sensor and a Kalman filter are used, is for instance described in [7] and [8]. The paper is organized as follows. Section II presents the model of a synchronous motor using the Park transform. Section III is devoted to the explanation of the Kalman filter which is used to approximate the derivative of the controller. Simulations and conclusions close the paper.

2. Model of a Synchronous Motor

Many models are known in the related literature concerning PMSM since the introduction of PMSM, the two-axis dq-model, obtained using Park dq-transformation is the most widely used in variable speed PMSM drive control applications [3] and [9]. The Park dq-transformation is a coordinate transformation that converts the three-phase stationary variables into variables in a rotating coordinate system. The dq-model is considered in this work.

$$\begin{bmatrix} u_d(t) \\ u_q(t) \\ u_0(t) \end{bmatrix} = \begin{bmatrix} \frac{2\sin(\omega_{el}t)}{3} & \frac{2\sin(\omega_{el}-2\pi/3)}{3} & \frac{2\sin(\omega_{el}+2\pi/3)}{3} \\ \frac{2\cos(\omega_{el}t)}{3} & \frac{2\cos(\omega_{el}-2\pi/3)}{3} & \frac{2\cos(\omega_{el}+2\pi/3)}{3} \\ \frac{1}{3} & \frac{1}{3} & \frac{1}{3} \end{bmatrix} \begin{bmatrix} u_a(t) \\ u_b(t) \\ u_c(t) \end{bmatrix}, \tag{1}$$

$$\begin{bmatrix} i_d(t) \\ i_q(t) \\ i_0(t) \end{bmatrix} = \begin{bmatrix} \frac{2\cos(\omega_{el}t)}{3} & \frac{2\cos(\omega_{el}-2\pi/3)}{3} & \frac{2\cos(\omega_{el}+2\pi/3)}{3} \\ \frac{-2\sin(\omega_{el}t)}{3} & \frac{-2\sin(\omega_{el}-2\pi/3)}{3} & \frac{-2\sin(\omega_{el}+2\pi/3)}{3} \\ \frac{1}{3} & \frac{1}{3} & \frac{1}{3} \end{bmatrix} \begin{bmatrix} i_a(t) \\ i_b(t) \\ i_c(t) \end{bmatrix}. \tag{2}$$

The dynamic model of the synchronous motor in dq-coordinates can be represented as follows:

$$\begin{bmatrix} \frac{di_d(t)}{dt} \\ \frac{di_q(t)}{dt} \end{bmatrix} = \begin{bmatrix} -\frac{R_s}{L_d} & \frac{L_q}{L_d}\omega_{el}(t) \\ -\frac{R_s}{L_q} & -\frac{L_d}{L_q}\omega_{el}(t) \end{bmatrix} \begin{bmatrix} i_d(t) \\ i_q(t) \end{bmatrix} + \begin{bmatrix} \frac{1}{L_d} & 0 \\ 0 & \frac{1}{L_q} \end{bmatrix} \begin{bmatrix} u_d(t) \\ u_q(t) \end{bmatrix} - \begin{bmatrix} 0 \\ \frac{\Phi\omega_{el}(t)}{L_q} \end{bmatrix}, \tag{3}$$

and

$$M_m = \frac{3}{2}p\Big\{\Phi i_q(t) + (L_d - L_q)i_d(t)i_q(t)\Big\}. \tag{4}$$

In (3) and (4), $i_d(t)$, $i_q(t)$, $u_d(t)$ and $u_q(t)$ are the dq-components of the stator currents and voltages in synchronously rotating rotor reference frame, $\omega_{el}(t)$ is the rotor electrical angular speed, the parameters R_s, L_d, L_q, Φ and p are the stator resistance, d-axis and q-axis inductance, the amplitude of the permanent magnet flux linkage, and p is the number of couples of permanent magnets, respectively. At the end, M_m indicates the motor torque. Considering an isotropic motor with $L_d \simeq L_q = L_{dq}$, it follows:

$$\begin{bmatrix} \frac{di_d(t)}{dt} \\ \frac{di_q(t)}{dt} \end{bmatrix} = \begin{bmatrix} -\frac{R_s}{L_{dq}} & \omega_{el}(t) \\ -\frac{R_s}{L_{dq}} & \omega_{el}(t) \end{bmatrix} \begin{bmatrix} i_d(t) \\ i_q(t) \end{bmatrix} + \begin{bmatrix} \frac{1}{L_{dq}} & 0 \\ 0 & \frac{1}{L_{dq}} \end{bmatrix} \begin{bmatrix} u_d(t) \\ u_q(t) \end{bmatrix} - \begin{bmatrix} 0 \\ \frac{\Phi\omega_{el}(t)}{L_{dq}} \end{bmatrix}, \quad (5)$$

and

$$M_m = \frac{3}{2} p\Phi i_q(t), \quad (6)$$

with the following movement equation:

$$M_m - M_w = J\frac{d\omega_{mec}(t)}{dt}, \quad (7)$$

where $p\omega_{mech}(t) = \omega_{el}(t)$ and M_w is an unknown mechanical load.

3. Using Kalman Filter to Approximate a Derivative Controller

The idea to control the system is to use the scheme reported in Fig. 1 in which a PID control structure is reported. In particular, it is possible to see the insertion of a Kalman filter which estimates the derivative of the velocity error using a model consisting of a polynomial of the first order. The objective is to avoid a direct derivative which generates noise and spikes due to the measured noise. Let us assume that the error in the velocity tracking can be expressed by a mathematical polynomial expression. Assuming, for sake of compactness, that the polynomial which we are considering is just a polynomial of the first order as follows:

$$e_v(t) = a_0 + a_1 t, \quad (8)$$

in which $e_v(t)$ represents the tracking velocity error. The following Kalman filter can be implemented in which at each sampling time constant parameters a_0 and a_1 should be calculated. The following state representation is obtained:

$$\begin{bmatrix} \dot{e}_v(t) \\ \dot{e}_a(t) \end{bmatrix} = \begin{bmatrix} 0 & 1 \\ 0 & 0 \end{bmatrix} \begin{bmatrix} e_v(t) \\ e_a(t) \end{bmatrix}. \quad (9)$$

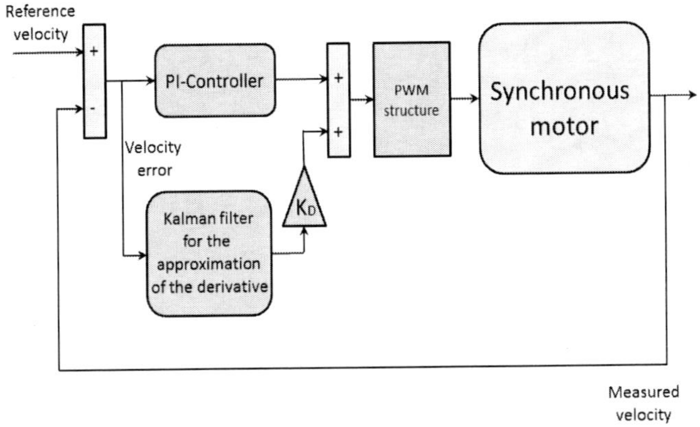

Figure 1. Complete control scheme.

It is to notice that $\dot{e}_v(t) = a_1$ represents the acceleration error (the derivative of the velocity error). In this sense, according to the following general notation, $\mathbf{e}(t) \in \mathbb{R}^2$, $u(t) \in \mathbb{R}$, and $y(t) \in \mathbb{R}$, then:

$$\begin{cases} \dot{\mathbf{e}}(t) = \mathbf{A}\mathbf{e}(t) + \mathbf{B}u(t) \\ y(t) = \mathbf{H}\mathbf{e}(t), \end{cases} \tag{10}$$

where

$$\dot{\mathbf{e}}(t) = \begin{bmatrix} \dot{e}_v(t) \\ \dot{e}_a(t) \end{bmatrix}, \tag{11}$$

$$\mathbf{A} = \begin{bmatrix} 0 & 1 \\ 0 & 0 \end{bmatrix}, \quad \mathbf{B} = \begin{bmatrix} 0 \\ 0 \end{bmatrix} \tag{12}$$

and $\mathbf{H} = [1 \;\; 0]$. Considering the discretisation of system of Eq. (10) the following discrete system is obtained:

$$\begin{cases} \dot{\mathbf{e}}(k/k-1) = \mathbf{A}_d\mathbf{e}(k-1/k-1) + \mathbf{B}_d u(k-1) + \mathbf{R}_w \\ y(k/k) = \mathbf{H}\mathbf{e}(k/k) + \zeta, \end{cases} \tag{13}$$

where $\mathbf{R}_w$ is the process noise covariance matrix and ζ is the measurement noise covariance. $\mathbf{A}_d$ and $\mathbf{B}_d$ represent the discrete forms of the matrix $\mathbf{A}$ and $\mathbf{B}$ of

Eq. (12) respectively. In particular, if T_s is the sampling time and Forward Euler Method is applied, the following matrices are obtained:

$$\mathbf{A}_d = \begin{bmatrix} 1 & T_s \\ 0 & 1 \end{bmatrix}, \quad \mathbf{B}_d = \begin{bmatrix} 0 \\ 0 \end{bmatrix}. \tag{14}$$

The a priori predicted state is

$$\mathbf{e}(k/k-1) = \mathbf{A}_d \mathbf{e}(k-1/k-1) + \mathbf{B}_d u(k-1), \tag{15}$$

and the a priori predicted covariance matrix is

$$\mathbf{P}(k/k-1) = \mathbf{A}_d \mathbf{P}(k-1/k-1) + \mathbf{P}(k/k-1)\mathbf{A}_d^T + \mathbf{R}_w, \tag{16}$$

The following equations state the correction (a posteriori prediction) of the Kalman filter:

$$\begin{aligned} \mathbf{K}(k) &= \mathbf{P}(k/k-1)\mathbf{H}_s^T \big(\mathbf{H}\mathbf{P}(k/k-1)\mathbf{H}^T + \zeta\big)^{-1} \\ \mathbf{e}(k/k) &= \mathbf{e}(k/k-1) + \mathbf{K}(k)\big(y(k) - \mathbf{H}\mathbf{e}(k/k-1)\big) \\ \mathbf{P}(k/k) &= \mathbf{P}(k/k-1) - \mathbf{K}(k)\mathbf{H}\mathbf{P}(k/k-1), \end{aligned} \tag{17}$$

where $\mathbf{K}(k)$ is the Kalman gain.

Remark 1. *It is to notice that matrix $\mathbf{R}_w$ (process covariance noise) consists of the following structure:*

$$\mathbf{R}_w = \begin{bmatrix} 0 & 0 \\ 0 & R_{2,2} \end{bmatrix}, \tag{18}$$

in which $R_{2,2}$ states a squared variance. The reason of this structure is that the first equation corresponding to matrix $\mathbf{A}$ of (12) is a definition of the velocity and in this sense no uncertainty should be set. The second equation is an equation which considers a comfortable condition (the derivative of the velocity error equals zero) but of course it is not true. In this case, an uncertainty variable must be set. According to our experience a very wide system needs a very wide uncertainty variable which must be set. □

4. Simulations

In this section simulated results are shown. Results have been performed using a special stand with a 58-kW traction PMSM. The stand consists of a PMSM, a tram wheel and a continuous rail. The PMSM is a prototype for low floor trams. The PMSM parameters are: nominal power 58 kW, nominal torque 852 Nm, nominal speed 650 rpm, nominal phase current 122 A, nominal input voltage 230 V and number of poles 44. The model parameters are: $R = 0.08723$ Ohm, $L_{dq} = Ld = Lq = 0.8$ mH, $\Phi = 0.167$ Wb. The engine has a nominal power 55 kW, a nominal voltage 380 V. Figure 2 shows how the tracking needs a filter of the derivative action. In fact, the left part of Fig. 2 presents the tracking obtained using a filtered PID controller and using a polynomial Kalman filter approach. In the right part of Fig. 2 also a simulation using a non-filtered derivative action is used in which an instability is shown. In particular, the following filtered PID controller is adopted:

$$G_{PID}(s) = K_p s + K_I \frac{1}{s} + K_D \frac{N}{1 + Ns}, \tag{19}$$

where with $G_{PID}(s)$ the transfer function on the complex s variable is indicated, K_P, K_I and K_D indicate the proportional, integral, and derivative coefficients of the controller and N indicates the filter coefficient. The results shown in Fig. 2 are obtained using $N = 100$, $K_P = 10$, $K_I = 1000$, $K_D = 100000$ and for the Kalman filter matrix

$$\mathbf{R}_w = \begin{bmatrix} 0 & 0 \\ 0 & 10000 \end{bmatrix}, \tag{20}$$

and $\zeta = 0.1$ starting with initial value of the covariance matrix which equals zero. Figure 3 shows the controlling switching signal using filtered PID (left) and using Kalman filter (right) and Fig. 4 shows the PWM signal using filtered PID (left) and using Kalman filter (right). It is to notice that using the filtered PID approach a lower switching frequency is obtained with a similar value of the tracking error.

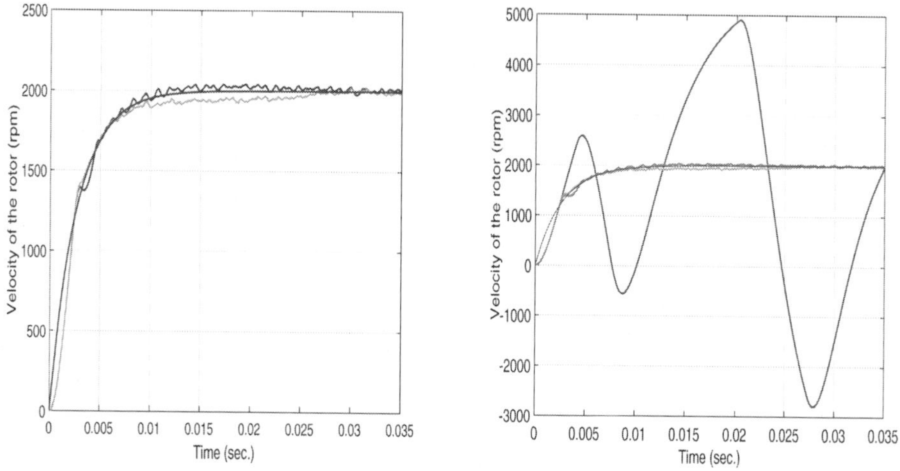

Figure 2. On the left: tracking obtained with filtered PID (blue) and with polynomial Kalman filter (yellow). On the right: a non-filtered derivative action is used.

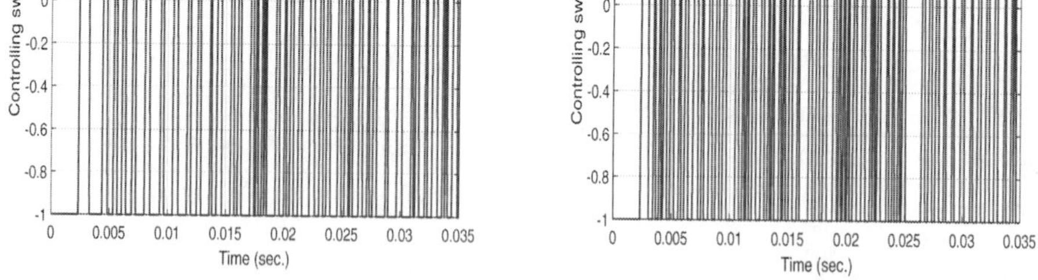

Figure 3. Controlling switching signal using filtered PID (left) and using Kalman filter (right).

Conclusion

This paper presented a PID control scheme for a synchronous motor with permanent magnets. A polynomial Kalman filter is used to approximate its derivative

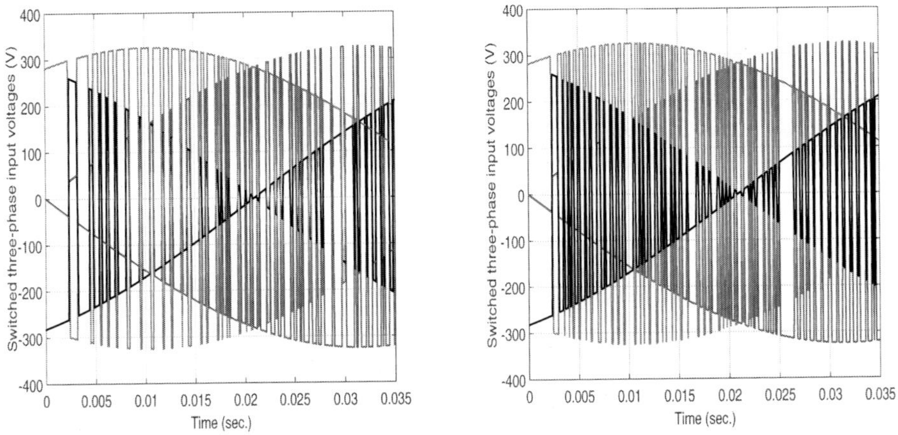

Figure 4. PWM signal using filtered PID (left) and using Kalma filter (right).

actions to avoid disturbance and spikes due to the differentiation. Future work can include possible higher order of the Kalman filter to be able to realise a D^2 controller. Moreover, a parameter optimisation of the PID coefficients can be considered too.

References

[1] M.A. Rahman and P. Zhou. Analysis of brushless permanent magnet synchronous motors. *IEEE Transactions on Industrial Electronics*, 43(2):256–267, 1996.

[2] M. Ooshima, A. Chiba, A. Rahman, and T. Fukao. An improved control method of buried-type IPM bearingless motors considering magnetic saturation and magnetic pull variation. *IEEE Transactions on Energy Conversion*, 19(3):569–575, 2004.

[3] M.A. Rahman, D.M. Vilathgamuwa, M.N. Uddin, and K.-J. Tseng. Nonlinear control of interior permanent magnet synchronous motor. *IEEE Transactions on Industry Applications*, 39(2):408–416, 2003.

[4] C. Tai, A. Stubbs, and T.-C. Tsao. Modeling and control design of an electromagnetic engine valve. In *Proceedings of the American control conference*, volume 4, pages 2890–2895, Arlington, VA, 2001.

[5] P. Mercorelli. Robust feedback linearization using an adaptive PD regulator for a sensorless control of a throttle valve. *Mechatronics a journal of IFAC*, 19(8):1334–1345, 2009.

[6] P. Mercorelli. An adaptive and optimized switching observer for sensorless control of an electromagnetic valve actuator in camless internal combustion engines. *Asian Journal of Control*, 16(4):959–973, 2014.

[7] P. Mercorelli. A two-stage sliding-mode high-gain observer to reduce uncertainties and disturbances effects for sensorless control in automotive applications. *IEEE Transactions on Industrial Electronics*, 62(9):5929–5940, 2015.

[8] P. Mercorelli. A hysteresis hybrid extended Kalman filter as an observer for sensorless valve control in camless internal combustion engines. *IEEE Transactions on Industry Applications*, 48(6):1940–1949, 2012.

[9] D.A. Khaburi and M. Shahnazari. Parameters identification of permanent magnet synchronous machine in vector control. In *Proceedings of the* 10th *European Conference on Power Electronics and Applications, EPE 2003*, page 660, Toulouse (France), 2003.

RELATED NOVA PUBLICATIONS

ROBUST H∞ FUZZY CONTROL OF NONLINEAR MARKOVIAN JUMP SYSTEMS: LMI-BASED DESIGN*

Wudhichai Assawinchaichote, Ph.D.†
Department of Electronic and Telecommunication Engineering,
Faculty of Engineering, King Mongkut's
University of Technology Thonburi,
Bangkok, Thailand

In this chapter, the problem of a robust $H\infty$ fuzzy control design for a class of nonlinear Markovin jump systems via LMI-based approach is examined. Based on Takagi-Sugeno fuzzy model, the robust $H\infty$ fuzzy controller is obtained for a nonlinear Markovian jumps system. Then, a set of sufficient conditions of the fuzzy controller for such a system is given to guarantee the closed-loop system asymptotic stability and the pre-prescribed $H\infty$ performance level. To show the effectiveness of the design results, finally a numerical example is provided.

---

* The full version of this chapter can be found in *Fuzzy Control Systems: Design, Analysis and Performance Evaluation*, edited by Wendy Santos, published by Nova Science Publishers, Inc, New York, 2016.

† E-mail address: wudhichai.asa@kmutt.ac.th.

LMIs Approach Based Robust H∞ Fuzzy Controller Design For Nonlinear Time-Varying Delay Systems*

Wudhichai Assawinchaichote, Ph.D. †
Department of Electronic and Telecommunication Engineering,
Faculty of Engineering, King Mongkut's
University of Technology Thonburi,
Bangkok, Thailand

In this chapter, the problem of designing a robust *H∞* fuzzy controller for nonlinear time-varying delay systems is considered. Based on a Takagi-Sugeno (TS) fuzzy model and linear matrix inequalities (LMIs) approach, the robust *H∞* fuzzy controller is obtained for such a system. A sufficient condition of the proposed controller is given to ensure the closed-loop system asymptotic stability and the prescribed *H∞* performance level. Finally, in this chapter, a numerical example is provided to illustrate the effectiveness of the design results.

Robust Control Approach to Option Pricing: The Uniqueness Theorem‡

Pierre Bernhard[1] and Naïma El Farouq[2]
[1] INRIA Sophia Antipolis-méditerranée, France
[2] University Blaise Pascal, Clermont-Ferrand, France

In this article, we sketch the overall context and prove the uniqueness sought. Notice, however, that the present proof does not account for the discontinuous

---

* The full version of this chapter can be found in *Control Theory: Perspectives, Applications and Developments*, edited by Francisco Miranda, published by Nova Science Publishers, Inc, New York, 2015.

† E-mail address: wudhichai.asa@kmutt.ac.th.

‡ The full version of this chapter can be found in *Game Theory and Applications. Volume 16*, edited by Leon Petrosjan and Vladimir Mazalov, published by Nova Science Publishers, Inc, New York, 2013.

payment digital option, while that of [3] can be extended to that case, thanks to the concept of barrier.

STABILIZATION, REGULATION AND ROBUST CONTROL OF UNCERTAIN PROCESSES AND PARAMETERS IN POROUS MEDIUM SYSTEMS WITH APPLICATIONS*

Aziz Belmiloudi†
Institut National des Sciences Appliquées de Rennes (INSA),
Institut de Recherche MAthématique de Rennes (IRMAR),
Rennes, France

This chapter investigates a robust control approach of uncertain complex systems in porous medium described by a class of degenerate nonlinear partial differential equations. The concept consists in setting the problem in the worst-case disturbances which leads to the game theory in which the controls and the disturbances (which destabilize the dynamical behavior of the system) play antagonistic roles. The aim of the approach is then to control these instabilities, either by acting on certain parameters to maintain the system in a desired state, or by calculating the limit of these parameters before the system becomes unstable (details of our approach are reported in Belmiloudi's book 2008).

A review of some practical applications and examples (which are mainly focused on the optimization of oil production and groundwater extraction) are presented. The dynamical models occurring in porous media are predicted by using some generalized uncertain evolution nonlinear porous medium equations with nonlinear Robin boundary conditions (with and without coupling with generalized Darcy's law).

The control and robust stabilization problems are formulated, in different situations, in order to reconstitute simultaneously, different parameters, data functions and sources/sinks appearing in the state system and in the boundary conditions, and to control and stabilize desired states provided by e.g., measurements (as pressure, production rates, time average of the moisture, etc).

* The full version of this chapter can be found in *Focus on Porous Media Research*, edited by Changying Zhao, published by Nova Science Publishers, Inc, New York, 2013.

†E-mail addresses: Aziz.Belmiloudi@insa-rennes.fr; Aziz.belmiloudi@math.cnrs.fr.

This work includes results concerning the existence of optimal solutions, sensitivity problems, adjoint problems, necessary optimality conditions (necessary to develop numerical optimization methods) and optimization problems. Applications to some problematics as optimal management of underground resources, control of oil reservoir production, etc, are presented; the case when data are measured in only some points in space-time domain, and the case when the objective is to find the best locations of a finite number of wells, which are drilled, are also considered.

BIBLIOGRAPHY

Advanced neural network-based computational schemes for robust fault diagnosis

LCCN	2013943920
Type of material	Book
Personal name	Mrugalski, Marcin, author.
Main title	Advanced neural network-based computational schemes for robust fault diagnosis / Marcin Mrugalski.
Published/Produced	Cham: Springer, [2014] ©2014
Description	xviii, 182 pages: illustrations; 25 cm.
Links	Publisher description https://www.loc.gov/catdir/enhancements/fy1610/2013943920-d.html Table of contents only https://www.loc.gov/catdir/enhancements/fy1610/2013943920-t.html
ISBN	331901546X 9783319015460
LC classification	QA76.87 .M78 2014
Summary	The present book is devoted to problems of adaptation of artificial neural networks to robust fault diagnosis schemes. It presents neural networks-based modeling and estimation techniques used for designing robust fault diagnosis schemes for non-linear dynamic systems. A part of the book focuses on fundamental issues such as architectures of dynamic neural networks, methods for designing of neural networks

	and fault diagnosis schemes as well as the importance of robustness. The book is of a tutorial value and can be perceived as a good starting point for the new-comers to this field. The book is also devoted to advanced schemes of description of neural model uncertainty. In particular, the methods of computation of neural networks uncertainty with robust parameter estimation are presented. Moreover, a novel approach for system identification with the state-space GMDH neural network is delivered. All the concepts described in this book are illustrated by both simple academic illustrative examples and practical applications.
Contents	Introduction -- Designing of dynamic neural networks -- Estimation methods in training of ANNs for robust fault diagnosis -- MLP in robust fault detection of static non-linear systems -- GMDH networks in robust fault detection of dynamic non-linear systems -- State-space GMDH networks for actuator robust FDI -- Conclusions and future research directions.
Subjects	Neural networks (Computer science) Fault location (Engineering) Robust control. Fault location (Engineering) Neural networks (Computer science) Robust control.
Notes	Includes bibliographical references (pages 169-179) and index.
Series	Studies in Computational Intelligence, 1860-949X; Volume 510 Studies in computational intelligence; v. 510.

Advances in computer science and information engineering. Volume 1

LCCN	2012937219
Type of material	Book
Meeting name	CSIE (Conference) (2012: Zhengzhou, China)
Main title	Advances in computer science and information engineering. Volume 1 / David Jin and Sally Lin (eds.).
Published/Produced	Berlin; New York: Springer, [2012]
Description	xix, 726 pages: illustrations; 24 cm.

ISBN	9783642301254 3642301258
LC classification	QA75.5 .C75 2012
Related names	Jin, David. edt Lin, Sally. edt
Contents	Design on architecture of internet of things / GuiPing Dai and Yong Wang -- XML-based structural representing method for information of things in internet of things / GuiPing Dai and Yong Wang -- A QoS-aware web service orchestration engine based on actors / GuiPing Dai and Yong Wang -- Emergency management application based on spatial information cloud service / Ken Chen, Fang Miao and WenHui Yang -- Research on complex networks evolution model / JinJun Xie -- Using ASP.NET to develop intelligent online ordering system in e-commerce / Jing Li -- Passive energy-dissipation control for aqueduct structural under earthquake based on SIMULINK / Liang Huang, Yujie Hou and Bo Wang -- Research on psychological characteristics of undergraduates and construction of study style for electronic-information-type specialties / Liqun Huang, Fulai Liu, Hongfei Mo and Xiang Cheng -- Research on self-organization Mechanism of discipline cluster system / Changhui Yang and Fengjuan Cao -- Study on clustering algorithms of wireless self-organized network / Haitao Wang, Lihua Song and Hui Chen -- Study on speech control of turning movements of the multifunctional nursing bed / Yingchun Zhong and Xing Hua -- Robust control for a mixed leg mechanism four-legged walking robot / ChuangFeng Huai, XueYan Jia and PingAn Liu -- Pedometer algorithm research based-Matlab / Bin Huang and Xinhui Wu -- Multi-user detection based on EM gradient algorithm / Jinlong Xian and Zhi Liu -- Multi-user detection based on the ECM iterative algorithm in Gaussian noise / Jinlong Xian and Zhi Liu -- Research on metamorphic testing for oracle problem of integer bugs / Yi Yao, Song Huang and Mengyu Ji -- Application of wireless

distributed sensor networks in subway energy saving systems / Jun Fan, Bing Xu and Peng Sun -- The Model of face recognition in video surveillance based on cloud computing / Shijun Yi, Xiaoping Jing, Jing Zhu, Jinshu Zhu and Heping Cheng -- A Research on the difference, equilibrium and evolution of the wage in inter-province: an analysis based on the equilibrium theory / Junbo Xue and Zheng Wang -- Study on the deployment process of low attitude interception net / Xiaofei Zhou, Yi Jiang and Yongyuan Li -- Construction and study of computer Agongqiang corpus / Guofu Yin -- A Semantic context-based access control model for pervasive computing environments / HaiBo Shen -- On the mean value of an arithmetical function / MingShun Yang and YaJuan Tu -- A 1-D image random scrambling algorithm based on wavelet transform / Qiudong Sun, Yunfeng Xue, Xueyi Chang and Yongping Qiu -- A Delay-guaranteed two-level polling model / Zheng Guan and Dongfeng Zhao -- E-commerce personalized recommendation system design based on multi-agent / Ya Luo -- Codebook perturbation based limited feedback precoding MU-MIMO system / Hongliang Fu, Beibei Zhang, Huawei Tao, Yong Tao and Zheng Luo -- Compressed sensing image de-noising method based on regularization of wavelet decomposition / Hongliang Fu, Huawei Tao, Beibei Zhang and Jing Lu -- The Optimal antenna array group for a novel QOSTBC decode algorithm / Yong Wang and Hui Li -- The Application of semantic vocabulary in building intelligent search system / HongSheng Xu and GuoFang Kuang -- Different characteristics of MCP and PAB in electricity market by evolutionary game theory / Xian Huang and Jialei Tan -- Study on the stable conditions of 3 x 3 symmetrical evolutionary games / Xian Huang -- The Parameter matching study of hybrid time-share drive electric vehicle on a high mobility vehicle / Yan Sun, Shishun Zhu, Sujun Luo, Jianwen Li and Peng Ye -- An Improved multiple attributes decision making based on ideal solution and project / HouXing Tang and WeiFeng Lu -- Unequal error protection for image transmission based on power water-filling policy over MIMO channels / Wu-ping Zhang, Xiao-rong Jing, Jun Li, Zu-fan Zhang and Qian-bin Chen -- The Delay of state detection in wireless monitoring system based on ZigBee / Renshu Wang, Yan Bai and Jianjun Zhao -- A Rough set approach to online customer's review mining / Wei Wang, Huan Yuan, Ye Chen, Ping Yang and Leilei Ni et al. -- Cluster center initialization using hierarchical two-division of a data set along each dimension / Guang Hui Chen -- The Design and implementation of a driving simulation system based on virtual reality technology / Hongyu Lai, Wei Huang, Peng Wei and Jingchang Pan -- A Novel PCA-GRNN flow pattern identification algorithm for electrical resistance tomography system / Yanjun Zhang and Yu Chen -- A Temperature difference determination method of infrared diagnostic for petrochemical equipment / Hua Lai, Yongwang Tang, Na Liu and Jianlin Mao -- Key technologies in digital library ontology / Qiong Su -- Factor analysis on M-learning effect and strategy based on information technology / Lin Hu, Honghua Xu, Shuangzi Sun and Chunfeng Xu -- Design and exploitation of energy MIS based on data integration / QingHua Pang and TianTian Lu -- An Improved incremental learning approach based on SVM model for network data stream / Na Sun and YanFeng Guo -- Improved region local binary patterns for image retrieval / Xiaosheng Wu and Junding Sun -- A New WSN-based routing optimization in smart distribution grid / GangJun Gong, XingChuan Wang and JuanXi Zhu -- Patent evaluation method of automotive parts manufacturing enterprise based on analytic hierarchy process and fuzzy matter-element method / Jun Chen, He Huang, He Lu, Jian Hu and Gangyan Li -- Application of improved genetic algorithm in logistics transportation / Jixian Xiao and Bing Lu -- Study on establishing public traffic network of Qianjiang City / Haijiang Ye -- Vehicle routing problem with soft time windows / Jixian Xiao and Bing Lu -- Research on monitoring data processing and fractional predictive method of the phreatic line / Xiaoxia Wang, Fengbao Yang, Linna Ji and Qingying Fan -- High-speed train adaptive control based on multiple models / Yonghua Zhou, Chao Mi and Yaqin Yin -- Generalized simulation models of vehicle movements for road and railway traffic using cellular automata with braking reference distance / Yonghua Zhou, Chao Mi and Yaqin Yin -- Multi-objective test paper evaluation in the process of composing with computer / Yan Li, Jiqiang Tang and Min Fu -- Breakout prediction system based on combined neural network in continuous casting / Xin Jin, Tingzhi Ren, Xueliang Shi, Rong Jin and Dawei Liu -- Prediction of house trade based on enthusiasm and purchasing power of buyers / Miao Xu and Qi Meng -- A 3D acoustic temperature field reconstruction algorithm based on Tikhonov regularization / Hua Yan, Kun Li and ShanHui Wang -- Intelligent identification of high rock-filled embankment elastic-plastic model parameters and application / Chengzhong Yang, Xianda Xu and Shufang Wang -- Distributed detection approach for wormhole attack in wireless sensor networks / Haofei Wang and Tongqiang Li -- Routing optimization design of wireless ad hoc network based on electric energy data acquire system / GangJun Gong, JuanXi Zhu, XingChuan Wang and Chen Xiong -- Prediction research based on improved intelligent algorithm model / Mingjun Wang and Shuxian Deng -- The Research of information search system based on formal concept analysis theory / GuoFang Kuang and HongSheng Xu -- Correlation analysis among international soybean futures markets based on D-vine pair-copula method / Jianbao Chen, Ting Yang and Huobiao Zhou -- The Research for the algorithm of digital image watermark based on decomposition of wavelet packet / ZhenLei Lv and HaiZhou Ma -- A Technology and implementation of embedding watermark to digital pictures dynamically / HaiZhou Ma and ZhenLei Lv -- Practice of bringing up mechatronics all-round innovation talents with measure, quality and standardization characteristics / Xiaolu Li and Bi Sun -- Research on frequency assignment in battlefield based on adaptive genetic algorithm / Libiao Tong, Hongjun Zhang, Wenjun Lu and Jie Zhou -- Research on model of circuit fault classification based on rough sets and SVM / Fu Yu, Zheng Zhi-song and Wu Xiao-ping -- Research on model of circuit fault diagnosis based on Bayesian inference / Fu Yu, Zheng Zhi-song and Wu Xiao-ping -- Research on the alkalinity of Sintering process based on LS-SVM algorithms / Rui Wang, Ai-min Wang and Qiang Song -- Leveling accuracy research based on Trimble 5700 GPS antenna / Tianzi Li, Huabin Chai and Yuanyuan Xu -- Feature learning based multi-scale wavelet analysis for textural image segmentation / Jing Fan -- Magneto-electric generator based on automotive vibration energy / Jian Hu, Jian Sun and Yanfei Zhou -- Handwriting digit recognition based on fractal edge feature and BP neural net / Yingchao Zhang, Tailei Liu and Xiaoling Ye -- Software development of crude oil storage system energy consumption analysis and comprehensive optimization / Jian Zhao, Yang Liu, Hang Dong and Xueqing Sun -- A Strategy of genetic operations based on schema / Fachao Li, Jie Yang and Chenxia Jin.

Subjects Computer science--Congresses.
Software engineering--Congresses.
Computer simulation--Congresses.

Notes International conference proceedings.
Includes bibliographical references and author index.

Series Advances in intelligent and soft computing, 1867-5662; 168
Advances in intelligent and soft computing; 168.

Applied control systems design

LCCN	2012936365
Type of material	Book
Personal name	Mahmoud, Magdi S.
Main title	Applied control systems design / Magdi S. Mahmoud, Yuanqing Xia.
Published/Created	London; New York: Springer, c2012.
Description	xxi, 560 p.: ill. (some col.); 24 cm.
ISBN	9781447128786 (alk. paper) 1447128788 (alk. paper) 9781447128793 (e-ISBN)
LC classification	TJ213 .M26843 2012
Related names	Xia, Yuanqing.
Contents	Introduction -- Some industrial systems -- System identification methods -- Applications I -- Nominal control design -- Applications II -- Robust control design -- Adaptive control -- Appendix.
Subjects	Automatic control.
Notes	Includes bibliographical references and index.

Attractive ellipsoids in robust control

LCCN	2014946581
Type of material	Book
Personal name	Poznyak, Alexander.
Main title	Attractive ellipsoids in robust control / Alexander Poznyak.
Published/Produced	New York: Springer, 2014.
Links	Table of contents only http://www.loc.gov/catdir/enhancements/fy1412/2014946581-t.html Publisher description http://www.loc.gov/catdir/enhancements/fy1412/2014946581-d.html Contributor biographical information http://www.loc.gov/catdir/enhancements/fy1412/2014946581-b.html
ISBN	9783319092096

Challenges and paradigms in applied robust control

LCCN	2016937445
Type of material	Book
Main title	Challenges and paradigms in applied robust control / [edited by] Beatrice Adamsen.
Edition	1st edition.
Published/Produced	Valley Cottage, NY: Scitus Academics, 2016.
ISBN	9781681172743 (hardcover: alk. paper)

Feedback control for computer systems

LCCN	2014498854
Type of material	Book
Personal name	Janert, Philipp K., author.
Main title	Feedback control for computer systems / Philipp K. Janert.
Edition	First edition.
Published/Produced	Beijing; Sebastopol, CA: O'Reilly, 2013. ©2014
Description	xvii, 308 pages: illustrations (some color); 23 cm
ISBN	9781449361693 (pbk.) 1449361692 (pbk.)
LC classification	TJ216 .J38 2013
Contents	Machine generated contents note: pt. I Foundations -- 1. Why Feedback? An Invitation -- A Hands-On Example -- Hoping for the Best -- Establishing Control -- Adding It Up -- Summary -- Code to Play With -- 2. Feedback Systems -- Systems and Signals -- Tracking Error and Corrective Action -- Stability, Performance, Accuracy -- The Setpoint -- Uncertainty and Change -- Feedback and Feedforward -- Feedback and Enterprise Systems -- Code to Play With -- 3. System Dynamics -- Lags and Delays -- Forced Response and Free Response -- Transient Response and Steady-State Response -- Dynamics in the Physical World and in the Virtual World -- Dynamics and Memory -- The Importance of Lags and Delays for Feedback Loops -- Avoiding Delays -- Theory and Practice -- Code to Play With -- 4. Controllers -- Block Diagrams -- On/Off Control -- Proportional Control --

Why Proportional Control Is Not Enough -- Integral Control -- Integral Control Changes the Dynamics -- Integral Control Can Generate a Constant Offset -- Derivative Control -- Problems with Derivative Control -- The Three-Term or PID Controller -- Code to Play With -- 5. Identifying Input and Output Signals -- Control Input and Output -- Directionality of the Input/Output Relation -- Examples -- Thermal Control 1 Heating -- Item Cache -- Server Scaling -- Controlling Supply and Demand by Dynamic Pricing -- Thermal Control 2 Cooling -- Criteria for Selecting Control Signals -- For Control Inputs -- For Control Outputs -- A Note on Multidimensional Systems -- 6. Review and Outlook -- The Feedback Idea -- Iteration -- Process Knowledge -- Avoiding Instability -- The Setpoint -- pt. II Practice -- 7. Theory Preview -- Frequency Representation -- The Transfer Function -- Block-Diagram Algebra -- PID Controllers -- Poles of the Transfer Function -- Process Models -- 8. Measuring the Transfer Function -- Static Input/Output Relation: The Process Characteristic -- Practical Considerations -- Dynamic Response to a Step Input: The Process Reaction Curve -- Practical Aspects -- Process Models -- Self-Regulating Process -- Accumulating Process -- Self-Regulating Process with Oscillation -- Non-Minimum Phase System -- Other Methods of System Identification -- 9. PID Tuning -- Tuning Objectives -- General Effect of Changes to Controller Parameters -- Ziegler-Nichols Tuning -- Semi-Analytical Tuning Methods -- Practical Aspects -- A Closer Look at Controller Tuning Formulas -- 10. Implementation Issues -- Actuator Saturation and Integrator Windup -- Preventing Integrator Windup -- Setpoint Changes and Integrator Preloading -- Smoothing the Derivative Term -- Choosing a Sampling Interval -- Variants of the PID Controller -- Incremental Form -- Error Feedback Versus Output Feedback -- The General Linear Digital Controller -- Nonlinear Controllers -- Error-Square and

Gap Controllers -- Simulating Floating-Point Output -- Categorical Output -- 11. Common Feedback Architectures -- Changing Operating Conditions: Gain Scheduling -- Gain Scheduling for Mildly Nonlinear Systems -- Large Disturbances: Feedforward -- Fast and Slow Dynamics: Nested or "Cascade" Control -- Systems Involving Delays: The Smith Predictor -- pt. III Case Studies -- 12. Exploring Control Systems Through Simulation -- The Case Studies -- Modeling Time -- Control Time -- Simulation Time -- The Simulation Framework -- Components -- Plants and Systems -- Controllers -- Actuators and Filters -- Convenience Functions for Standard Loops -- Generating Graphical Output -- 13. Case Study: Cache Hit Rate -- Defining Components -- Cache Misses as Manufacturing Defects -- Measuring System Characteristics -- Controller Tuning -- Simulation Code -- 14. Case Study: Ad Delivery -- The Situation -- Measuring System Characteristics -- Establishing Control -- Improving Performance -- Variations -- Cumulative Goal -- Gain Scheduling -- Integrator Preloading -- Weekend Effects -- Simulation Code -- 15. Case Study: Scaling Server Instances -- The Situation -- Measuring and Tuning -- Reaching 100 Percent With a Nonstandard Controller -- Dealing with Latency -- Simulation Code -- 16. Case Study: Waiting-Queue Control -- On the Nature of Queues and Buffers -- The Architecture -- Setup and Tuning -- Derivative Control to the Rescue -- Controller Alternatives -- Simulation Code -- 17. Case Study: Cooling Fan Speed -- The Situation -- The Model -- Tuning and Commissioning -- Closed-Loop Performance -- Simulation Code -- 18. Case Study: Controlling Memory Consumption in a Game Engine -- The Situation -- Problem Analysis -- Architecture Alternatives -- A Nontraditional Loop Arrangement -- A Traditional Loop with Logarithms -- Results -- Simulation Code -- 19. Case Study Wrap-Up -- Simple Controllers, Simple Loops -- Measuring and Tuning --

Staying in Control -- Dealing with Noise -- pt. IV Theory -- 20. The Transfer Function -- Differential Equations -- Laplace Transforms -- Properties of the Laplace Transform -- Using the Laplace Transform to Solve Differential Equations -- A Worked Example -- The Transfer Function -- Worked Example: Step Response -- Worked Example: Ramp Input -- The Harmonic Oscillator -- What If the Differential Equation Is Not Known? -- 21. Block-Diagram Algebra and the Feedback Equation -- Composite Systems -- The Feedback Equation -- An Alternative Derivation of the Feedback Equation -- Block-Diagram Algebra -- Limitations and Importance of Transfer Function Methods -- 22. PID Controllers -- The Transfer Function of the PID Controller -- The Canonical Form of the PID Controller -- The General Controller -- Proportional Droop Revisited -- A Worked Example -- 23. Poles and Zeros -- Structure of a Transfer Function -- Effect of Poles and Zeros -- Special Cases and Additional Details -- Pole Positions and Response Patterns -- Dominant Poles -- Pole Placement -- What to Do About Delays -- 24. Root Locus Techniques -- Construction of Root Locus Diagrams -- Root Locus or "Evans" Rules -- Angle and Magnitude Criteria -- Practical Issues -- Examples -- Simple Lag with a P Controller -- Simple Lag with a PI Controller -- 25. Frequency Response and the Bode Plot -- Frequency Response -- Frequency Response in the Physical World -- Frequency Response for Transfer Functions -- A Worked Example -- The Bode Plot -- A Criterion for Marginal Stability -- Other Graphical Techniques -- 26. Topics Beyond This Book -- Discrete-Time Modeling and the z-Transform -- State-Space Methods -- Robust Control -- Optimal Control -- Mathematical Control Theory -- pt. V Appendices.

Subjects	Feedback control systems. Control theory. Control theory.

	Feedback control systems.
Notes	Includes bibliographical references (pages 297-299) and index.

H [infinity] robust designs and their applications to control, signal processing, communication, systems and synthetic biology: an integrated course for engineering, mathematics, and bioscience

LCCN	2015013346
Type of material	Book
Personal name	Chen, Bor-Sen.
Main title	H [infinity] robust designs and their applications to control, signal processing, communication, systems and synthetic biology: an integrated course for engineering, mathematics, and bioscience / Bor-Sen Chen and Chia-Chou Wu, editors.
Published/Produced	New York: Nova Publishers, [2016]
Description	x, 441 pages; 26 cm.
ISBN	9781634824927 (hardcover)
LC classification	TJ217.2 .C483 2016
Related names	Chen, Bor-Sen, editor.
	Wu, Chia-Chou, editor.
Subjects	Robust control--Mathematics.
	H [infinity symbol] control.
Notes	In title "[infinity]" appears as the infinity symbol.
	Includes bibliographical references and index.
Series	Systems biology--theory, techniques and applications

Linear robust control

LCCN	2012010587
Type of material	Book
Personal name	Green, Michael.
Main title	Linear robust control / Michael Green, David J.N. Limebeer.
Edition	Dover ed.
Published/Created	Mineola, N.Y.: Dover Publications, Inc., 2012.
Description	xv, 538 p.: ill.; 24 cm.
ISBN	9780486488363 (pbk.)
	0486488365 (pbk.)
LC classification	TJ213 .G755 2012

Related names	Limebeer, David J. N.
Summary	"Recent years have witnessed enormous strides in the field of robust control of dynamical systems -- unfortunately, many of these developments have only been accessible to a small group of experts. In this text for students and control engineers, the authors examines all of these advances, providing an in-depth and exhaustive examination of modern optimal and robust control. "-- Provided by publisher.
Subjects	Linear control systems. Linear systems. SCIENCE / System Theory.
Notes	Originally published: Englewood Cliffs, N.J.: Prentice Hall, c1995. Includes bibliographical references (p. 507-523) and index.

LMIs in control systems: analysis, design and applications

LCCN	2013008019
Type of material	Book
Personal name	Duan, Guangren.
Main title	LMIs in control systems: analysis, design and applications / Guang-Ren Duan, Hai-Hua Yu.
Published/Produced	Boca Raton: CRC Press, Taylor & Francis Group, [2013]
Description	xxix, 453 pages; 24 cm
Links	Cover image http://jacketsearch.tandf.co.uk/common/jackets/covers/websmall/978146658/9781466582996.jpg
ISBN	9781466582996 (hardback)
LC classification	QA402.3 .D7973 2013
Summary	"Preface In recent years linear matrix inequalities (LMIs) have emerged as a powerful tool in the field of control systems analysis and design. Many problems such as state feedback synthesis, robustness analysis and design, H2 and H? control, can all be reduced to convex or quasi-convex problems that involve LMIs. The most attractive feature of LMIs lies in the fact that many problems in systems and control can be reduced to LMI

problems, which can be solved both efficiently and numerical reliably. Very efficient algorithms to solve linear matrix inequalities have been developed and can be readily used through the mature Matlab LMI toolbox or the open source software packages YALMIP and CVX. LMIs today has been a real technique. With the help of the LMI techniques in control systems analysis and design, more and more theoretical and practical applications can now be solved, many of which might not be otherwise solved by traditional methods. Goal of the book The goal of this book is to provide a textbook for graduate and senior undergraduate courses in the field of control systems analysis and design, which contains the basic but systematic knowledge about LMIs and LMI-based control systems analysis and design. Boyd et al. (1994), which is the first book on LMIs, certainly has performed a pioneering function in the field, but limited by time it fails to catch up with the ever fast emerging new theories and techniques in LMIs. Yu (2002) is a book on LMI approach to robust control, which is suitable to be used as a textbook, but the Chinese language limits its usage worldwide"--Provided by publisher.

Subjects	Control theory. Matrix inequalities. Mathematical optimization. Mathematics / Applied Technology & Engineering / Electrical Technology & Engineering / Engineering (General)
Notes	Includes bibliographical references (pages 433-449) and index.

Loop-shaping robust control

LCCN	2013936315
Type of material	Book
Personal name	Feyel, Philippe.
Main title	Loop-shaping robust control / Philippe Feyel.
Published/Created	London: ISTE; Hoboken, NJ: Wiley, 2013.
Description	xxiv, 261 p.: ill.; 24 cm.

ISBN	9781848214651 (hbk.) 1848214650 (hbk.)
LC classification	TJ217.2 .F49 2013
Subjects	Robust control. Mathematical optimization. Mathematical optimization Robust control.
Notes	Includes bibliographical references (p. [255]-258) and index.
Series	Automation-control and industrial engineering series Automation-control and industrial engineering series.

Mathematical methods in robust control of linear stochastic systems

LCCN	2013945412
Type of material	Book
Personal name	Drăgan, Vasile.
Main title	Mathematical methods in robust control of linear stochastic systems / Adrian-Mihail Stoica.
Edition	Second edition.
Published/Produced	New York: Springer, [2013] ©2013
Description	xv, 442 pages; 24 cm
Links	Publisher description http://www.loc.gov/catdir/enhancements/fy1404/2013945412-d.html Table of contents only http://www.loc.gov/catdir/enhancements/fy1404/2013945412-t.html
ISBN	9781461486626 (acid-free paper)
LC classification	TJ217.2 .D74 2013
Related names	Morozan, Toader. Stoica, Adrian.
Subjects	Robust control--Mathematical models. Linear systems. Stochastic systems.
Notes	Includes bibliographical references.

Optimal and robust control: advanced topics with MATLAB́

LCCN	2011046677
Type of material	Book
Personal name	Fortuna, L. (Luigi), 1953-
Main title	Optimal and robust control: advanced topics with MATLAB́ / edited by Luigi Fortuna, Mattia Frasca.
Published/Created	Boca Raton, FL: CRC Press, c2012.
Description	xviii, 233 p.: ill.; 25 cm.
ISBN	9781466501911 (hardback: acid-free paper)
LC classification	TJ217.2 .F67 2012
Related names	Frasca, Mattia.
Summary	"The scope of this book is to give to people with the knowledge of the main concepts of automatic control and signals and systems analysis a self-contained resource for advanced techniques both in linear system theory and in control design. This is achieved by including selected theoretical backgrounds, many numerical exercises and MATLAB examples. In this work we offer a complete and easy-to-read handbook of advanced topics in automatic control including the LQR and the H [infinity symbol] approach. Moreover our task has been to take advanced concepts of modeling analysis and to present the essential items of the new technique of LMI in order to show how it can be considered as a unifying tool for system analysis and controller design. The robustness property of the closed control loop is the guideline of the book. The text deals with advanced automatic control techniques with particular attention to their robustness"--Preface.
Subjects	Robust control--Mathematical models. Linear systems. Automatic control. Technology & Engineering / Electrical. Technology & Engineering / Electronics / Microelectronics. Technology & Engineering / Mechanical.
Notes	Includes bibliographical references (p. 217-222) and index.

Optimal and robust scheduling for networked control systems

LCCN	2012050720
Type of material	Book
Personal name	Longo, Stefano.
Main title	Optimal and robust scheduling for networked control systems / Stefano Longo, Tingli Su, Guido Herrmann, Phil Barber.
Published/Produced	Boca Raton, FL: CRC Press, Taylor & Francis Group, [2013] ©2013
Description	xxxi, 245 pages; 24 cm
Links	Cover image http://jacketsearch.tandf.co.uk/common/jackets/covers/websmall/978146656/9781466569546.jpg
ISBN	9781466569546 (hardback) 1466569549 (hardback)
LC classification	TJ217.2 .L64 2013
Related names	Su, Tingli. Herrmann, Guido. Barber, Phil.
Summary	"This book offers a tool for optimal/robust control system integration via communication networks. This work is shaped around examples and relevant standards of the automotive industry for those examples but the concepts are readily extendable to any application that uses deterministic communication protocols between system components (avionic systems, robots, etc.). The underlying idea is to use the rigorous tools from optimal and robust multivariable control theory to solve the industrial scheduling problem in a transparent manner"-- Provided by publisher.
Subjects	Robust control. Automatic control. Computer networks. Technology & Engineering / Electrical Technology & Engineering / Mechanical
Notes	Includes bibliographical references (pages 231-242) and index.

Series	Automation and control engineering: a series of reference books and textbooks

Parallel robots: mechanics and control

LCCN	2012040401
Type of material	Book
Personal name	Taghirad, Hamid.
Main title	Parallel robots: mechanics and control / Hamid Taghirad.
Published/Created	Boca Raton: Taylor & Francis, c2013.
Description	xiii, 519 p.: ill.; 26 cm.
ISBN	9781466555761 (hardback)
LC classification	TJ211.4152 .T34 2013
Summary	"Using a systematic approach and illustrative examples, this book examines dynamic and control for parallel robotics through three methods--Newton-Euler, principle of virtual work, and Lagrange formulations. It also presents a method for formulating dynamics equations into closed form, leading to more tractable dynamics matrices and easier verification. The author presents a concise overview of kinematic, static, and singularity analysis that should be required in any text on parallel robotics. The book also contains numerous case studies to empower readers to implement controllers in challenging situations. Contains all analysis and design tools required to create a parallel robot. Combines coverage of kinematics, dynamics, and control. Presents numerous illustrative case studies and many challenging chapter-end problems. Discusses cable-driven robots. Covers advanced topics such as dynamics model verification, redundancy resolution, and adaptive and robust control"-- Provided by publisher.
Subjects	Parallel robots. Technology & Engineering / Electrical Technology & Engineering / Electronics / General Technology & Engineering / Robotics
Notes	Includes bibliographical references (p. 501-509) and index.

Randomized algorithms for analysis and control of uncertain systems: with applications

LCCN	2012951683
Type of material	Book
Personal name	Tempo, R., author.
Main title	Randomized algorithms for analysis and control of uncertain systems: with applications / Roberto Tempo, Giuseppe Calafiore, Fabrizio Dabbene.
Edition	Second Edition.
Published/Produced	London: Springer, [2013]
Description	xxiii, 357 pages: illustrations (some color); 24 cm.
ISBN	9781447146094 (alk. paper) 1447146093 (alk. paper)
LC classification	QA402 .T43 2013
Related names	Calafiore, Giuseppe, 1969- author. Dabbene, Fabrizio, 1968- author.
Summary	"The presence of uncertainty in a system description has always been a critical issue in control. The main objective of Randomized Algorithms for Analysis and Control of Uncertain Systems, with Applications (Second Edition) is to introduce the reader to the fundamentals of probabilistic methods in the analysis and design of systems subject to deterministic and stochastic uncertainty. The approach propounded by this text guarantees a reduction in the computational complexity of classical control algorithms and in the conservativeness of standard robust control techniques. The second edition has been thoroughly updated to reflect recent research and new applications with chapters on statistical learning theory, sequential methods for control and the scenario approach being completely rewritten."-- Publisher's website.
Contents	1. Overview -- 2. Elements of probability theory -- 3. Uncertain linear systems -- 4. Linear robust control design -- 5. Limits of the robustness paradigm -- 6. Probabilistic methods for uncertain systems -- 7. Monte Carlo methods -- 8. Probability inequalities -- 9. Statistical learning theory -- 10. Randomized algorithms in systems and control -- 11. Sequential

	methods for probabilistic design -- 12. Scenario approach to probabilistic design -- 13. Learning-based probabilistic design -- 14. Random number and variate generation -- 15. Statistical theory of random vectors -- 16. Vector randomization methods -- 17. Statistical theory of random matrices -- 18. Matrix randomization methods -- 19. Applications of randomized algorithms.
Subjects	Control theory. System analysis. Stochastic processes. Algorithms.
Notes	Includes bibliographical references (pages 337-352) and index.
Series	Communications and control engineering, 0178-5354

Robot motion and control 2011

LCCN	2011941381
Type of material	Book
Meeting name	International Workshop on Robot Motion and Control (8th: 2011: Bukowy Dworek, Poland)
Main title	Robot motion and control 2011 / Krzystof R. Kozłowski (ed.).
Published/Produced	London; New York: Springer, [2012] ©2012
Description	xxiv, 418 pages: illustrations (some color); 24 cm.
Links	Contributor biographical information http://www.loc.gov/catdir/enhancements/fy1318/2011941381-b.html Publisher description http://www.loc.gov/catdir/enhancements/fy1318/2011941381-d.html Table of contents only http://www.loc.gov/catdir/enhancements/fy1318/2011941381-t.html
ISBN	9781447123422 1447123425
LC classification	TJ211.4 .I58 2011
Related names	Kozłowski, Krzysztof.
Contents	Symmetries in observer design: review of some recent results and applications to EKF-based SLAM -- Constraint control of mobile manipulators -- Control

methods for wheeeler-the hypermobile robot -- Tracking control strategy for the standard N-trailer mobile robot-a geometrically motivated approach -- Robust control of differentially driven mobile platforms -- Mobile system for non destructive testing of weld joints via time of flight diffraction (TOFD) technique -- Task-priority motion planning of wheeled mobile robots subject to slipping -- First steps toward automatically generating bipedal robotic walking from human data -- Dynamic position/force controller of a four degree-of-freedom robotic leg -- Perception-based motion planning for a walking robot in rugged terrain -- Improving accuracy of local maps with active haptic sensing -- Postural equilibrium in two-legged locomotion -- Generalized predictive control of parallel robots -- Specification of a multi-agent robot-based reconfigurable fixture control system -- Indirect linearization concept through the forward model-based control system -- Research oriented motor controllers for robotic applications -- Efficient and simple noise filtering for stabilization turning of a novel version of a model reference adaptive controller -- Real-time estimation and adaptive control of flexible joint space manipulators -- Visual servo control admitting joint range of motion maximally -- Optimal extended Jacobian inverse kinematics algorithm with application to attitude control of robotic manipulators.

Active disturbance rejection control for a flexible-joint manipulator -- Collision free coverage with multiple agents -- Manipulating ergodic bodies through gentle guidance -- Practical efficiency evaluation of a nilpotent approximation for driftless nonholonomic systems -- Obstacle avoidance and trajectory tracking using fluid-based approach in 2D space -- Motion planning of nonholonomic systems-nondeterministic endogenous configuration space approach -- Generation of time optimal trajectories of an autonomous airship -- Formation flight control scheme for unmanned aerial vehicles -- Comparison of two-

	and four-engine propulsion structures of airship -- Dynamic simulations of free-floating space robots -- Planning motion of manipulators with local manipulability optimization -- A comparison study of discontinuous control algorithms for a three-link nonholonomic manipulator -- Development of knee joint robot with flexion, extension and rotation movements-experiments on imitation of knee joint movement of healthy and disable persons -- Exercise programming and control system of the leg rehabilitation robot RRH1.
Subjects	Robots--Motion--Congresses. Robots--Control systems--Congresses.
Notes	"Thirty one original works have been selected out of 39 papers submitted to the Eighth International Workshop on Robot Motion and Control (RoMoCo'11), Bukowy Dworek, Poland, June 15-17, 2011"--Page v. Includes bibliographical references and indexes.
Additional formats	Also issued online.
Series	Lecture notes in control and information sciences, 0170-8643; 422 Lecture notes in control and information sciences; 422.

Robust and adaptive control: with aerospace applications

LCCN	2012951336
Type of material	Book
Personal name	Lavretsky, Eugene.
Main title	Robust and adaptive control: with aerospace applications / Eugene Lavretsky, Kevin A. Wise.
Published/Produced	London; New York: Springer, [2013] ©2013
Description	xvii, 454 pages: illustrations (some color); 24 cm.
ISBN	9781447143956 (alk. paper) 1447143957 (alk. paper)
LC classification	TJ217 .L38 2013
Related names	Wise, Kevin A.
Contents	Part 1. Robust Control -- Introduction -- Optimal Control and the Linear Quadratic Regulator --

	Command Tracking and the Robust Servomechanism -- State Feedback H Optimal Control -- Frequency Domain Analysis -- Output Feedback Control -- Part 2. Robust Adaptive Control -- Direct Model Reference Adaptive Control: Motivation and Introduction -- Lyapunov Stability of Motion -- State Feedback Direct Model Reference Adaptive Control -- Model Reference Adaptive Control with Integral Feedback Connections -- Robust Adaptive Control -- Approximation-Based Adaptive Control -- Adaptive Control with Improved Transient Dynamics -- Robust and Adaptive Control with Output Feedback.
Subjects	Robust control. Adaptive control systems. Adaptive control systems. Robust control.
Notes	Includes bibliographical references and index.
Series	Advanced textbooks in control and signal processing, 1439-2232 Advanced textbooks in control and signal processing.

Robust control and linear parameter varying approaches: application to vehicle dynamics

LCCN	2012955946
Type of material	Book
Meeting name	International Summer School in Automatic Control (32nd: 2011: Grenoble, France)
Main title	Robust control and linear parameter varying approaches: application to vehicle dynamics / Olivier Sename, Péter Gáspár, and Jozsef Bokor (eds.).
Published/Produced	Heidelberg: Springer, [2013].
Description	xviii, 395 pages: illustrations (some color); 24 cm.
ISBN	9783642361098 (pbk.: acid-free paper) 3642361099 (pbk.: acid-free paper)
LC classification	TJ217.2 .R615 2013
Related names	Sename, Olivier Gáspár, Péter Bokor, J. (József), 1948-
Subjects	Robust control--Congresses.

	Motor vehicles--Dynamics--Congresses.
Notes	Proceeding of the 32nd International Summer School in Automatic Control held in Grenoble, France, September 12-16, 2011. International conference proceedings. Includes bibliographical references and index.
Series	Lecture notes in control and information sciences, 0170-8643; 437 Lecture notes in control and information sciences; 437.

Robust control design with Matlab

LCCN	2013938042
Type of material	Book
Personal name	Gu, Da-Wei.
Main title	Robust control design with Matlab / Da-Wei Gu, Petko H. Petkov, Mihail M. Konstantinov.
Edition	2nd edition.
Published/Produced	New York: Springer, 2013.
ISBN	9781447146810 (softcover: alk. paper)
Series	Advanced textbooks in control and signal processing

Robust control for grid voltage stability: high penetration of renewable energy.

LCCN	2014941696
Type of material	Book
Main title	Robust control for grid voltage stability: high penetration of renewable energy.
Published/Produced	New York, N.Y.: Springer Berlin Heidelberg, 2014.
Links	Contributor biographical information http://www.loc.gov/catdir/enhancements/fy1411/2014941696-b.html Publisher description http://www.loc.gov/catdir/enhancements/fy1411/2014941696-d.html Table of contents only http://www.loc.gov/catdir/enhancements/fy1411/2014941696-t.html

Robust control of uncertain dynamic systems

LCCN	2013954668
Type of material	Book
Personal name	Yedavalli, Rama K.
Main title	Robust control of uncertain dynamic systems / Rama K. Yedavalli.
Published/Produced	New York: Springer, 2013.
Links	Contributor biographical information http://www.loc.gov/catdir/enhancements/fy1405/2013954668-b.html Publisher description http://www.loc.gov/catdir/enhancements/fy1405/2013954668-d.html Table of contents only http://www.loc.gov/catdir/enhancements/fy1405/2013954668-t.html
ISBN	9781461491316

Robust mechanism design: the role of private information and higher order beliefs

LCCN	2012005125
Type of material	Book
Personal name	Bergemann, Dirk.
Main title	Robust mechanism design: the role of private information and higher order beliefs / by Dirk Bergemann & Stephen Morris.
Edition	1st ed.
Published/Created	Singapore; Hackensack, NJ: World Scientific Pub., 2012.
Description	xvii, 452 p.: ill.; 24 cm.
ISBN	9789814374583 (hbk.) 981437458X 9789814374590 9814374598
LC classification	TJ217.2 .B47 2012
Related names	Morris, Stephen.
Contents	Foreword -- Introduction -- Robust mechanism design -- Ex post implementation -- Robust implementation in direct mechanisms -- Robust implementation in general mechanisms -- The role of the common prior in robust implementation -- An ascending auction for interdependent values: uniqueness and robustness to

	strategic uncertainty -- Robust virtual implementation -- Multidimensional private value auctions -- The robustness of robust implementation -- Rationalizable implementation -- Pricing without priors.
Subjects	Robust control.
	Prices.
Notes	Includes bibliographical references and index.
Series	World scientific series in economic theory; 2

Robust methods for data reduction

LCCN	2016303296
Type of material	Book
Personal name	Farcomeni, Alessio, author.
Main title	Robust methods for data reduction / Alessio Farcomeni, Luca Greco.
Published/Produced	Boca Raton, FL: CRC Press, Taylor & Francis Group, [2015]
	©2015
Description	xxvii, 269 pages: illustrations; 24 cm
ISBN	9781466590625
	1466590629
LC classification	QA276.4 .F38 2015
Related names	Greco, Luca, author.
Contents	Section 1. Dimension reduction -- section 2. Sample reduction.
Subjects	Robust control.
	Data reduction--Computer programs.
	Dimension reduction (Statistics)
	Data Interpretation, Statistical.
	Data reduction--Computer programs.
	Dimension reduction (Statistics)
	Robust control.
	Datenanalyse.
	Datenkompression.
	R.
Notes	"A Chapman & Hall book."
	Includes bibliographical references (page 245-266) and index.

Sb-robustness of performance measures of control chart

LCCN	2013334628
Type of material	Book
Personal name	Laha, Arnab Kumar, author.
Main title	Sb-robustness of performance measures of control chart [microform] / Arnab Kumar Laha, Pravinda Raja.
Published/Produced	Ahmedabad, India: Indian Institute of Management, 2013.
Description	1 microfiche (19 leaves)
LC classification	Microfiche 2015/60012 (T)
Related names	Indian Institute of Management, Ahmedabad.
Subjects	Robust control--Mathematical model.
Notes	Includes bibliographical references (leaves 16-19).
Reproduction no./Source	Library of Congress -- New Delhi Overseas Office
Additional formats	Microfiche. New Delhi: Library of Congress Office; Washington, D.C.: Library of Congress Photoduplication Service, 2013. 1 microfiche Master microform held by: DLC.
Series	Working paper; W.P. no. 2013-04-01

S-variable approach to LMI-based robust control.

LCCN	2014948741
Type of material	Book
Main title	S-variable approach to LMI-based robust control.
Published/Produced	New York: Springer, 2014.
Links	Publisher description http://www.loc.gov/catdir/enhancements/fy1412/2014948741-d.html Contributor biographical information http://www.loc.gov/catdir/enhancements/fy1412/2014948741-b.html Table of contents only http://www.loc.gov/catdir/enhancements/fy1412/2014948741-t.html
ISBN	9781447166054

INDEX

D

E

F

G

H

I

J

K

L

M

N

O

P

Q

R

S

T

U

V

W

X